HANDBOOK FOR TEACHING AND LEARNING IN GEOGRAPHY

1975

Also by Dylan Jones:

James Bond Style (with Lindy Hemming)

These Foolish Things: A Memoir

Loaded: The Unexpurgated Oral History of the Velvet Underground

Magic: A Journal of Song (with Paul Weller)

Faster Than a Cannonball: 1995 and All That

Shiny & New: Ten Moments of Pop Genius that Defined the Eighties

Sweet Dreams: From Club Culture to Style Culture

The Wichita Lineman: Searching in the Sun for the World's Greatest Unfinished Song

David Bowie: A Life

London Sartorial

Manxiety

London Rules

Mr Mojo

Elvis Has Left the Building: The Day the King Died

The Eighties: One Day, One Decade

From the Ground Up: The U2 360 Tour

When Ziggy Played Guitar: David Bowie and Four Minutes that Shook the World

The Biographical Dictionary of Popular Music

Heroes (with David Bailey)

Cameron on Cameron: Conversations with Dylan Jones (with David Cameron)

Mr Jones' Rules for the Modern Man

iPod Therefore I Am: A Personal Journey Through Music

Sex, Power and Travel: 10 Years of Arena (ed.)

Meaty, Beaty, Big & Bouncy: Classic Rock and Pop Writing from Elvis to Oasis (ed.)

Ultra Lounge

True Brit: Paul Smith (with Paul Smith)

Jim Morrison: Dark Star

Haircults

The i-D Bible (co-ed.)

1975

The Year the World Forgot

DYLAN JONES

C

CONSTABLE

CONSTABLE

First published in Great Britain in 2025 by Constable

1 3 5 7 9 10 8 6 4 2

Copyright © Dylan Jones, 2025

The moral right of the author has been asserted.

A CIP catalogue record for this book
is available from the British Library.

ISBN: 978-1-40872-198-8 (hardback)

Typeset in Kepler Std by SX Composing DTP, Rayleigh, Essex
Printed and bound in Great Britain by Clays Ltd, Elcograf S.p.A.

Papers used by Constable are from well-managed forests
and other responsible sources.

FSC
www.fsc.org
MIX
Paper | Supporting
responsible forestry
FSC® C104740

Constable
An imprint of
Little, Brown Book Group
Carmelite House
50 Victoria Embankment
London EC4Y 0DZ

The authorised representative
in the EEA is
Hachette Ireland
8 Castlecourt Centre, Dublin 15,
D15 XTP3, Ireland
(email: info@hbgi.ie)

An Hachette UK Company
www.hachette.co.uk

www.littlebrown.co.uk

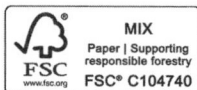

For Edie and Georgia

CONTENTS

Los Angeles: A British Fantasy

*'A change in the weather is sufficient to recreate the world
and ourselves' – Marcel Proust*

NINETEEN-SEVENTY-FOUR ENDED IN AN INCREDIBLY provident manner. On New Year's Eve, Mick Fleetwood asked Lindsey Buckingham and Stevie Nicks to officially join Fleetwood Mac, so creating not just one of the superpowered rock groups of the seventies, but also forging a sound and a sensibility that would start to peak six months later with the release of their first collaboration, the eponymous *Fleetwood Mac* album, a record that would help define the year as well as the decade. It was a concept album whose concept was Los Angeles, as never had a record sounded so Californian, so sumptuous, so golden (honestly, you almost expected the album to come complete with a pair of sunglasses and a poolside ice bucket).

In the same way that, more than 150 years ago, manifest destiny drove American pioneers westward – as hordes of speculators, migrants and would-be moguls staked claim to anything and everything before them as they pressed onwards to the Pacific Ocean – so during the late sixties and early seventies, Los Angeles became the geographic holy grail of American rock music. It didn't matter if you were an aspiring singer-songwriter like Joni Mitchell or Neil Young, an eager bunch of double-denim guitar players like the Eagles, or an old British blues band like Fleetwood Mac looking for rejuvenation, LA was where you came. It was the Electric Eden, Pacific Exotic, a landscape that never failed to charm.

'We all watched the sun set in the west every night of adolescence and thought someday about coming out here,' said the Eagles' Glenn Frey in the *LA Times*. 'It all seemed so romantic . . . The *Life* magazine articles about Golden Gate Park and the Sunset Strip . . . and the music of the Beach Boys, the Byrds, Buffalo Springfield. It was definitely the archetype of the most beautiful place in the world.'

While the city was still synonymous with Hollywood, with its gilded movie stars, Adobe-style mansions, palm trees and Pacific vistas, at the turn of the decade the Sunset Strip and the local canyons had been colonised by a new generation of migrating musicians who were drawn to the area because of its weather (naturally) and a burgeoning alternative lifestyle. Community. Music. Drugs. Suntanned limbs. And money. This was the land of blue jeans and mirrored shades, of cocaine and convertibles and freeways that went straight to the ocean. Like Vienna at the turn of the last century or Paris in the thirties, in California there was a sense of freedom in the air. It was an enchanted land.

The music industry had also started to shift west from New York, as the culture moved away from Brill Building pop to Laurel Canyon experimentation. The fantasy suggested that all you had to do was walk barefoot along the Sunset Strip and someone would come up to you and offer you a record contract. Sunset Boulevard stretched from the newest hardscrabble immigrants at one end to the oldest and richest immigrants at the other, a twenty-five-mile arrival desk. It was smack-bang in the middle where things were happening, and where dreams were coming true. The geography of the new scene featured the Troubadour, Barney's Beanery, the Laurel Canyon Country Store, Wonderland Avenue and a host of snazzy new recording studios opening up all over the city. By the mid-seventies, Los Angeles wasn't just an interesting musical hub, it was the centre of the entire industry. The Californian dream had become a reality. The denim had become designer. Musicians were the new movie stars, with the same mystique, the same gilded lifestyles. Here was seclusion and ostentation, the twin peaks of wealth and fame. In 1975, Los Angeles was still exotic, a testament to the power of ambition, a place where every palm tree looked like a reward, every freeway a route to success. LA was music city and every new arrival was a potential celestial cowboy. But these new arrivals weren't pioneers, they were settlers.

'The hippie version of freedom in the sixties was breaking down the Establishment,' said record producer Lou Adler. 'We were buying houses in Bel Air; we were becoming the Establishment.'

'When we came out of the sixties, when pop music had got to be such a revolutionary force, we kind of got nostalgic for the era before that, enamoured of old R&B records,' said Michael McDonald, who sang with two great examples of seventies stalwarts, Steely Dan and the Doobie Brothers. 'For us, that music was a kind of panacea to thinking, Well, has much actually changed, socially, after all that effort in the sixties? In the seventies, we kind of got lulled into an escapism of sorts. Our songs had more chord changes, they were harking back to a more romantic kind of musical experience. Steely Dan wanted to be Duke Ellington's band, you know?'

A friend once said Fleetwood Mac's Wikipedia page reads like a Russian novel, with new characters popping up before exiting in grim circumstances such as mental illness, alcoholism, adultery, a religious cult and romantic trauma. They're not wrong. The band was formed in 1967 in London by the guitarist Peter Green, who recruited drummer Mick Fleetwood and bassist John McVie. Honing a hip, blues-rock sound, they had commercial success with songs such as 'Black Magic Woman', 'Man of the World', 'Oh Well' and 'Albatross'. However, Green's use of LSD exacerbated his schizophrenia, causing him to quit the band in 1970. He was replaced by Christine Perfect soon after she married John McVie. Various other members came and went, rarely having much lasting impact.

Seeking a reinvention of sorts, in 1974 the band moved to Los Angeles, which was when Mick Fleetwood inadvertently bumped into Lindsey Buckingham at Sound City studio in the San Fernando Valley. He was checking out the studio when the producer played him a demo he'd been working on, just to demonstrate the sound. It turned out to be 'Frozen Love' from the *Buckingham Nicks* album; at that point

Buckingham put his head round the door, and the rest is rock and roll history. The couple radically altered the band's sound, adding a West Coast sheen that would quickly result in hit songs such as 'Over My Head', 'Say You Love Me', 'Rhiannon' and 'Landslide' – rich, luscious music that would help make the decade's mid-point one of the most important in the post-war narrative arc of pop.

In 1975, transatlantic travel was still an exotic experience, only available to the well-to-do. On 21 January 1970, Pan Am started flying passengers from New York to London aboard the very first commercial 747s. Spiral staircases connected two decks, widescreen movies were played in the cabins, and alcohol was available on tap: in the early seventies, the party never stopped. American Airlines removed sixty seats from its luxury fleet of 747s to build a lounge large enough to fit both a piano and a bar that served complimentary cocktails. Continental, meanwhile, equipped its planes with flying pubs, complete with arcade games and complimentary bottles of spirits. If you were a musician travelling from London to LA, you felt as though you deserved it. Pan Am boasted that hot meals were prepared simultaneously in four galley kitchens aboard each flight, and 'vibration-free' innovation meant airlines were able to deck out their cabins and triple-sized bathrooms with vases full of fresh-cut flowers. When you arrived on the West Coast you ended up somewhere that didn't just look futuristic, but felt genuinely alien.

And the Brits loved it. The allure of LA was a geographical manifestation of the US cultural hegemony at the time. The city represented the cultural uplands, an aesthetical dreamland which embraced the commodification of the rapidly developing counterculture.

Nineteen-seventy-five was the last year before the onslaught of punk, the last year in which the sophistication of seventies rock ruled imperiously, unencumbered by the arrival of the scrawny new broom. There is a myth that the long, dark days before punk were full of legions of Southern boogie bands and British prog rock groups, as the likes of Bachman-Turner Overdrive and ZZ Top, and Emerson Lake & Palmer and Jethro Tull roamed the lands, soiling the culture like fancy-dress petrol-pump attendants and university-educated orcs. Received wisdom had it that the immediate period before punk was littered with bloated old dinosaurs who squandered all genuine relationship with their audiences. Bands had grown too old, we were told, too fat, too rich and too pleased with themselves. They had emulated the movie stars of old.

Not strictly true. The mid-seventies was actually dense with extraordinarily sophisticated, mature rock music made by singers, songwriters and musicians who had no problem calling themselves artists. None at all. And the records they made aspired to artistic status: everyone was trying to make their own masterpiece and the sense of competitiveness had not been seen since the mid-sixties. Three-minute milquetoast pop singles had given way to concept albums, pop package tours had been supplanted by rock festivals, and rock in general had a renewed sense of ambition.

Everyone responsible for these records was aged between twenty-five (Steely Dan's Walter Becker, while Bruce Springsteen was twenty-six) and thirty-five (Smokey Robinson, with Bob Dylan, Chick Corea and George Clinton aged thirty-four). This was the generation the punks would rebel against, twenty- and thirty-somethings who probably already felt as though they'd lived several lifetimes. Neil Young seemed old even to his peers, and he was just thirty, the same age as Bryan Ferry, Bob Marley and Keith Jarrett. Many had different frames of reference, but all were boomers in one way or another. They may have achieved fame at different ages (Patti Smith was twenty-nine in 1975, whereas David Bowie was only twenty-eight and had been having hits since 1969), but they were all cut from the same cloth, often the same yard of denim. It always amused me that Brian Eno and Donna Summer were exactly the same age.

Radio changed in the seventies, as albums became more popular than singles. FM took over from AM, offering superior sound and not plagued by the same static and interference. Mirroring the move in quality, FM radio celebrated music of great length, depth and breadth, produced to an increasingly exalted level. The introduction of FM car radios also helped move the dial. A bigger bandwidth meant you could hear the detail in recordings that were becoming increasingly sophisticated. As twenty-four-track recordings started to become ubiquitous, so consumers wanted the kind of hardware and software that was going to do them justice. By 1975 you could take a mediocre idea and, via a production process that was tantamount to gentrification, turn it into something that was instantly revered and fetishised. This rush to professionalism and ego produced some genuine pillars of ambition, which would make 1975 one of the most important post-war years of all.

It was the apotheosis of adult pop, a year that was rich with near-masterpieces and *popularibus rebus*: *Blood on the Tracks* by Bob Dylan, *Nighthawks at the Diner* by Tom Waits, *The Who by Numbers* by the Who, *Still Crazy After all These Years* by Paul Simon, *Young Americans* by David Bowie, *The Köln Concert* by Keith Jarrett, *Siren* by Roxy Music, *Nils Lofgren* by Nils Lofgren, *Wish You Were Here* by Pink Floyd, *Born to Run* by Bruce Springsteen, *Another Green World* by Brian Eno, *The Hissing of Summer Lawns* by Joni Mitchell, *Captain Fantastic and the Brown Dirt Cowboy* by Elton John, *That's the Way of the World* by Earth, Wind and Fire, *There's No Place Like America Today* by Curtis Mayfield, *The Last Record Album* by Little Feat, *Pour Down Like Silver* by Richard and Linda Thompson, *Tomorrow Belongs to Me* by the Sensational Alex Harvey Band, *Marcus Garvey* by Burning Spear, *Pieces of the Sky* by Emmylou Harris, *Katy Lied* by Steely Dan, *One of These Nights* by the Eagles, *Chocolate City* by Parliament, *Black and Blue* by the Rolling Stones, *Why Can't We Be Friends?* by War, *Frampton* by Peter Frampton, *A Night at the Opera* by Queen, *The Original Soundtrack* by 10cc, *Dreaming My Dreams* by Waylon Jennings, *The Heat Is On* by the Isley Brothers, *Ruth is Stranger than Richard* by Robert Wyatt etc. This was civilisation, right here.

In 1975, the *New Musical Express* (then the great bastion of cultural criticism) were still calling the Who the greatest rock and roll band in the world, the greatest live act and still the best platform for adult insurrection (courtesy of Pete Townshend, of course). *The Who by*

Numbers was also an anomaly, as it was neither cobbled together (like their first three albums, *Odds & Sods* and even the magisterial *Who's Next*), nor was it a concept album like *Tommy* or *Quadrophenia*. And yet it was a colossus, a monster of a record that as much as anything was a status report on Townshend's relationship with his band, his industry and himself. When Townshend first shared the brutally honest, emotionally desolate songs that would comprise *The Who by Numbers*, Keith Moon walked over from his drum kit and gave the guitarist a massive hug. Townshend was growing up (he was thirty) and he was writing about it. The most important song on the record was 'How Many Friends', with lyrics that sounded as though they came straight from the therapist's couch. On the one hand, they concerned rock and roll's inability to change the world (one of Townshend's favourite topics), and on the other they reflected his worrying state of mind. The lyrics alluded to Townshend's conflicted sexual identity and distance from the rest of the group. As *Rolling Stone* so eloquently put it, here 'the protagonist is an aging, still successful rock star, staring drunkenly at the tube with a bottle of gin perched on his head, contemplating his career, his love for the music and his fear that it's all slipping away.'

Adult-Orientated Rock would soon become a pejorative, and yet here it was in all its thirty-something glory. Rock stars were writing about being older than they were when they started, about reaching an age they never properly contemplated. They were writing about what it felt to leave your teenage years behind. They were writing about the future, but not in the way they had before.

Van Morrison, another sixties star who had just hit thirty, was still basking in the adulation that had come his way after he'd released *Veedon Fleece* the year before. Having recorded *Astral Weeks* in 1969 and been swamped with praise for its originality and soulfulness, there had been an insidious narrative based around the idea that Morrison would never top it. But not only was every record he released after that a milestone of Celtic sophistication, *Veedon Fleece* was viewed by many to be even better than *Astral Weeks*. It was a record of such pastoral beauty it had put Albinoni, Chopin and Mendelssohn to shame. One critic described part of the record as the sound that grass makes when it's growing and if there was ever a musical equivalent of Walt Whitman, then this was it.

Morrison actually enjoyed getting old, as the very idea of being a pop star was anathema to him. He was genuinely enigmatic and went to great lengths to distance himself from the world of pop. He even wrote a letter to the Irish *Sunday Independent*, complaining about their constant references to Van Morrison 'the Rock Star', 'To call me a rock star is absurd,' he wrote, 'as anyone who has listened to my music will observe.'

Of course, the best albums of 1975 were driven by a belief that somehow what they were doing was of undoubted benefit to the consumer, but they were applauded for this, by critics and public alike. What rock fan wasn't going to feel empowered by listening to *Physical Graffiti*? Which wayward child of the suburbs wasn't going to be seduced (if only fleetingly) by Lou Reed's *Metal Machine Music*? What disco nut wasn't going to wallow in the futuristic hedonism of Donna Summer's *Love to Love You Baby*, the title track of which lasted an astonishing seventeen minutes and took up the whole first side? Which intellectual butterfly wasn't going to be seduced by this rush of extravagance?

Until 1975, the most celebrated reflection of the black urban experience had been the sweet social commentary of 1971's *What's Going On* by Marvin Gaye. *Chocolate City* by Parliament, George Clinton and Bootsy Collins's furious celebration of Washington DC, was a far more empowering statement, epitomised by the narrative in the title track: 'They still call it the White House, but that's a temporary condition.' The album was a didactic diatribe, an allegorical sidestep driven by rumbling funk, where the message really was in the music. Hypothesising what it might be like to have a black administration, the song assigns the following notables to positions of power – Muhammed Ali: president; Aretha Franklin: first lady; James Brown: veep; Richard Pryor: minister of education; Stevie Wonder: secretary of fine arts. Soon, the black experience would be better served by the emerging power of rap, but pre-punk, it was all about *Chocolate City*.

These records were magisterial, and while the world would soon be full of second division Joni Mitchells, Bob Marleys and Led Zeppelins, the whole point of them was that they couldn't be bettered. Who could realistically make a more sophisticated album than *The Hissing of Summer Lawns*? Or a more complex hard rock album than *Physical Graffiti*? Or a record as unimpeachable and as prescient as *Horses*?

'No one expected me,' said Patti Smith at the time. 'Everything awaited me,' she added, by way of explanation. And, beyond cocky, and convinced of her own importance, she meant it.

Here was the rock canon right before our eyes, a bunch of records that are increasingly recognised as the benchmarks not just of mid-seventies rock music, but of all time. There was nothing postmodern about these albums, either, nothing that nodded towards the contrary or the ironic. They were less about ideology and far more about extreme personal expression. Perhaps that's why so many of the great records of the period were American. To those of us here in the cold and the rain, the sunny climes of the US just seemed so colourful and alluring. The Americans certainly thought themselves as not tethered to any particular pop narrative. They were just doing what the hell they liked. America was still bathed in a mythic glow and, while Watergate and Vietnam had taken the shine off their imperial ideology, the country still seemed warm and expansive. This is what the philosopher Jean Baudrillard described as 'Astral America . . . Star-blasted, horizontally by the car, altitudinally by the plane, electronically by television, geologically by deserts . . . the power museum . . . for the whole world.'

This was a new Gilded Age, a time of great cultural prosperity but political corruption. If a mood board of the sixties looked like a fairground poster of psychedelic Victoriana as seen through a pair of rose-tinted glasses, its seventies' equivalent looked like an interiors shoot in a glossy magazine, probably featuring tubular steel lighting, cedar-toned furniture, geometric wallpaper and gargantuan, if perfectly manicured, palms. Levity might be suggested by some Hawaiian-style bamboo lettering. The only revolution that happened in the seventies was a lifestyle revolution. If the sixties were all about limitless possibilities, the seventies brought commodified culture for the bourgeoisie. If the sixties had been a glorious explosion of post-war expression, the seventies would try and turn it into hyperreal entertainment. The world was too scary a place to do anything else. Nevertheless, real life had a way of creeping its way into the culture regardless. Post-Watergate cynicism would start to cake the consciousness of the political and popular imagination, providing a dense, gritty texture that affected everything from music to cinema to literature, as well as the recently empowered genre of long-form journalism. Socially, culturally, geopolitically, the world was changing.

By the mid-seventies, pop media was changing too, driven in the US by titles such as *Rolling Stone* and *Creem,* and in the UK by the *New Musical Express,* the *Melody Maker* and *Sounds*, all of which foregrounded writing that emerged out of personal involvement, rather than the objectivity and fawning that was the bread and butter of traditional coverage of the entertainment industry. In this respect, the music press was echoing the alternative press of the sixties. By 1975 the mainstream had subsumed the underground, much like the music itself.

In many respects, the mid-seventies were an unapologetic flight into seriousness, a domain determined by talent as much as attitude. The album, meanwhile, was venerated as much as the modern novel. It was a unit of expertise, a measure of worth. And 1975 would be the zenith of what we would soon call 'classic rock'. The largest consumer demographic was all mostly the same age, between early teens and mid- to late-twenties. And while there was a greater distance between these two markers in the seventies, as a homogeneous mass they all acted broadly as one. You might have been a fan of Pink Floyd's moody and portentous *Wish You Were Here* at the age of fifteen, and you might see them in some gargantuan hall like Earls Court: the people in front of you in the expensive cowboy boots, Snoopy badges and butterfly-collared shirts weren't going to be over thirty. Not much, anyway. We were all experiencing this music for the first time, whether we were novices or seasoned customers. Music no longer had a single purpose, which made it generally more interesting. The broadly sketched counter cultural explosions of the sixties had shattered any idea of a common cause: individuality was now its own cause, and of course every individual had individual tastes. There was something for everyone! In 1975 there was something for a Smokey Robinson person, a Steely Dan person, a Bob Marley person. Cursory or obsessive, it was all here.

Generally, there was an attempt at respectability. Whether you were writing about the rich (Joni Mitchell), elaborately dissecting the modern marriage (Bob Dylan), using personal tragedy as a metaphor for political collapse (Neil Young) or channelling the hypnosis of social conditioning (Kraftwerk), you were producing work to be written about at length in the *New York Times* (if you were lucky). It wasn't enough to have a cover story in *Rolling Stone*; you also needed a three-thousand-word essay which compared your work to nineteenth-century French novelists or sixteenth-century Italian painters. In the minds of the people making them, as well as the people underwriting them, these albums were proper works of art.

The times deserved them. The US was just starting to come out of a recession caused by OPEC's oil crisis, heavy government spending on the Vietnam War, and a Wall Street stock crash. During the 1973 Arab-Israeli War, Arab members of the Organization of Petroleum Exporting Countries had imposed an embargo on the United States in retaliation for their decision to re-supply the Israeli military and to gain leverage in the post-war peace negotiations (Egypt would reopen the Suez canal in June). It was also the longest economic downturn in the US since the Great Depression. Unemployment peaked at 9 per cent, 2.3 million jobs were lost, and the dollar had an average inflation rate of 10.08 per cent per year between 1973 and 1975, producing a cumulative price increase of -17.47 per cent. Nevertheless, the American economy began to emerge from the recession during March 1975. Industrial production started to improve, the unemployment and inflation rates began to fall and consumer confidence even showed a slight increase.

At the end of April, the last few Americans still in South Vietnam were airlifted out of the country as Saigon fell to communist forces. American diplomats were on the front lines, organising what would be the most ambitious helicopter evacuation in history. North Vietnamese Colonel Bùi Tín, accepting the surrender of South Vietnam later in the day, remarked, 'You have nothing to fear; between Vietnamese there are no victors and no vanquished. Only the Americans have been defeated.'

The Vietnam War was the longest and most unpopular foreign war in US history and cost 58,000 American lives. As many as two million Vietnamese soldiers and civilians were killed. In light of his loss of political support and the near certainty that he would be impeached and removed from office because of the Watergate scandal (Watergate players John Mitchell, H. R. Haldeman and John D. Ehrlichman were sentenced to two-and-a-half to eight years for conspiracy and obstruction of justice) Richard Nixon resigned the presidency on 9 August 1974, the day after addressing the nation on television. The Nixons flew to their home in San Clemente, California, along with many of his staff. Every day he was at his desk by 7 a.m., surrounded by his press team, who had almost nothing to do. Some would say his successor in the White House, Gerald Ford, did much the same. Nineteen-seventy-four was tumultuous, but it also felt strangely vacant, an un-year. Elsewhere, the overthrow of Ethiopia's Haile Selassie in 1974 shook the Rastafari community, while his subsequent death in 1975 resulted in a crisis of faith for many practitioners. The year also witnessed the death of Franco, ending more than forty years of Spanish dictatorship.

Nineteen-seventy-five would also see the return of someone who had represented the very best of sixties' idealism, George Harrison. Not that anyone was interested, mind. While the two primary Beatles would both release substandard records – Paul McCartney and Wings's *Venus and Mars* and John Lennon's *Rock 'n' Roll* – it was Harrison's *Extra Texture* that was the real gem. The fact it was ignored shows the way in which the sixties were being finally laid to rest. This was Harrison's 'soul' record; sure, it was melancholic, but it was full of swing and melody. The album included 'This Guitar (Can't Keep From Crying)', which was both a sequel to Harrison's 1968 'While My Guitar Gently Weeps' and a rebuttal to *Rolling Stone*, whose savaging of his recent tour he would never forgive. Critics hated the album, framing it as Harrison's weakest, most slapdash work and confirmation he was in decline. They said he was 'burning out' and 'running out of gas'. Ironically, Harrison hadn't enjoyed making it as he hadn't enjoyed being in LA. He found it too dark and too cynical.

The city naturally had its flip side. The likes of Joni Mitchell, Gram Parsons, Jackson Browne and Crosby, Stills, Nash & Young had created a singer-songwriter culture that was based primarily in Laurel Canyon, a kind of resting place for the counter cultural, denim-clad wastrels who had moved to the city at the end of the sixties. For them – and the disenfranchised, rebellious youths who followed them – LA was a different place by the mid-seventies, a cynical city mired in cocaine. Linda Ronstadt, another member of the Hollywood country set, said cocaine 'made people deaf, it made people dead and it made people real obnoxious.' It was impossible to avoid, as the music industry appeared to be fuelled by the drug. No record company meeting was complete without someone getting out their coke spoon, no host would dream of holding a party without a coffee table full of white powder and it became the city's transactional drug of choice – for record contracts, media interviews, management deals, sex, everything.

Neil Bogart was the legendary head of Casablanca Records, who masterminded the careers of Kiss, Donna Summer and George Clinton's Parliament. He was known for his drug use, which for a time defined the company. He famously had a buzzer under his desk to control entry into his office. It was a necessary precaution, as every week a drug dealer known as 'Pock Face', with bad acne, would visit to take his orders. He needed privacy. 'It all went down on expense reports,' said one executive. 'An ounce of weed was, "A nice steak and a bottle of Bordeaux."' So much weed was smoked that the aroma permeated the rest of the building via the air conditioning. Cocaine was routinely snorted off desks, regardless of who was in the office at the time. 'Was it worse for drugs at Casablanca than at other record companies?' ponders Kiss's teetotal Gene Simmons, as published in *Classic Rock* in 2022. 'Yes, it was. I never approved of it. But the atmosphere was never mean-spirited, like it was at other companies.'

At the time it was difficult to think of Los Angeles as anything but a debaucherous sprawl, a zip code of barely disguised opportunity. Anyone in the music industry turning up at LAX from Heathrow or Gatwick was treated like visiting rock royalty, even songwriters.

'As soon as you made it through immigration there would be a limousine with blacked out windows waiting for you,' said one regular A&M Records writer, a Brit who spent much of 1975 in the air. 'Inside would be an extremely compliant young lady with a five-hundred-dollar smile who would proceed to open a jewel box full of cocaine and then give you a blow job as you were driven to the Sunset Marquis. There was nothing odd about it. The only strange thing would have been if it didn't happen. It was a perk of the job. Every record company was obsessed with keeping their talent, and woman and drugs were commodities.'

The hippies were now breadheads: on their massive 1974 tour, Crosby, Stills, Nash & Young were different men from the ones who had encouraged us all to 'Teach Your Children Well'; Stephen Stills told *Rolling Stone*'s Cameron Crowe that in terms of tours, 'We did one for the art and the music, one for the chicks. This one's for the cash.' It wasn't unusual to see crew members wearing the popular T-shirt, 'NO HEAD, NO BACKSTAGE PASS'. Sex was everywhere in 1975, especially at the infamous Troubadour, the legendary club on Santa Monica Boulevard owned by the equally legendary Doug Weston. The author Eve Babitz (who was once photographed playing chess naked with artist Marcel Duchamp) used to hang out at the bar. 'You can smell the semen out on the street,' she said, appalled at the libidinous nature of the club. One of the waitresses said, 'You had to wear a diaphragm just to walk through – it was enough to get you pregnant just standing there.'

At the top of the indulgence tree sat Led Zeppelin (the double-denim Vikings) and the Eagles (the celestial cowboys), with their personalised planes, imported cases of Château Lafite Rothschild, loosey-goosey groupies and mini-mountains of coke. The Eagles' hotel '3E' (or 'Third Encore') parties were Bacchanalian epics that will probably never be equalled. 'Led Zeppelin could argue with us, but I think we might have thrown the greatest travelling party of the seventies,' said Glenn Frey. Don Henley's enthusiasm for extracurricular activities made him something of a Hollywood legend. One particular incident at his California home, which featured five sex workers, a sex aid and prodigious amounts of cocaine, was excessive enough to feature in a book, the famous sexual exposé of La-La Land *You'll Never Make Love in This Town Again*. A few years later, Henley called paramedics to his home, where a sixteen-year-old girl was found naked and claiming she had overdosed on quaaludes and cocaine. She was arrested for prostitution, while a fifteen-year-old girl found in the house was arrested for being under the influence of drugs. Henley was arrested and subsequently charged for contributing to the delinquency of a minor. He pled no contest and was fined $2,500 and put on two years' probation.

The freewheeling sixties' culture had morphed into something far more strident, which in its own way was just as confrontational, a culture trying to capture the turbulence of the times. And Hollywood was feeding into that. The wider culture involved the still-unmatched dominance of US independent cinema, and a counterculture that was being subsumed into mainstream entertainment (as the Vietnam War ended, the sixties finally came to an end, ignoring the artificial calendrical finish five years earlier). This was a golden period of renewal in film with a roster of genuine auteurs: Martin Scorsese, Francis Ford Coppola, Robert Altman, Hal Ashby, Terrence Malick, Peter Bogdanovich, Roman Polanski, William Friedkin, Dennis Hopper, Bob Rafelson, Steven Spielberg and George Lucas who bestowed upon us such monumental works as *The Godfather, Nashville, Five Easy Pieces,*

Taxi Driver, Jaws, The Last Detail, Star Wars and *The Exorcist*, many of which reflected the societal upheavals of the times.

Warren Beatty's *Shampoo* was another film that caught the zeitgeist in 1975. Co-written by Beatty and Robert Towne, the movie follows a promiscuous Beverly Hills hairdresser on election day in 1968, as he juggles his relationships with various women ('As long as I can remember, when I see a pretty girl and I go after her and I make her, it's like I'm going to live for ever,' says his character). The film is ostensibly a satire focusing on the theme of sexual politics and late-sixties' sexual and social mores, and yet it couldn't be more seventies. After all, this is a film about a hairdresser; in the sixties, this was an occupation that, while you could squint and imagine a British version starring Michael Caine in his *Alfie* guise, would more than likely have been turned into a silly farce, a Swinging London sex-romp or faux-James Bond knock-off like the Matt Helm parodies. By 1975, being a Los Angeles hairdresser had genuine cachet; the phenomenally successful Hollywood producer Jon Peters got his break by cutting Barbra Streisand's locks, having established himself as a hairdresser who dyed women's pubic hair. At the time, Beatty was as famous for his prolific and unabashedly public love life as he was for being an ambitious cinematic auteur, and even though the film was a light-hearted goof on Beatty's movie-star persona, in a weird way it legitimised him. To amplify the echoes of Beatty's tabloid image, two of his real-life ex-girlfriends, Goldie Hawn and Julie Christie, were cast as two of his conquests. It didn't hurt the film at the box office, either, that one of Christie's lines – 'I want to suck his cock!' – became a cause célèbre (at a time when such language was still considered verboten by most mainstream media).

Miloš Forman's *One Flew Over the Cuckoo's Nest* was the second highest-grossing film of the year and was one of the high points of seventies cinema. Based on the novel by Ken Kesey, the movie stars Jack Nicholson as a new patient in an Oregon psychiatric hospital, alongside Louise Fletcher as a sadistic nurse. Originally announced in 1962 with Kirk Douglas starring, the film took thirteen years to develop. Released on 19 November, it's considered to be one of the greatest films ever made and was the second to win all five major Academy Awards (Best Picture, Best Actor, Best Actress, Best Director and Best Screenplay), following *It Happened One Night* in 1934. The *New York Times* said Nicholson slipped into his role with such easy grace that it was difficult to remember him in any other film.

Jaws premiered in the US on 20 June, when Americans were still reeling from Watergate and from the final death spiral in Vietnam. Here was an opportunity for escapism, one the public grabbed with both hands. The film not only changed the way we perceived swimming in the ocean, but it also changed for ever the film industry as we knew it – Hollywood, the summer and the notion of the seasonal blockbuster. Almost immediately it became the highest-grossing movie in history, a critical as well as popular success that was perceived as a new cultural phenomenon, focusing on fast-paced and exciting entertainment which would spark conversations beyond the cinema as well as repeated viewings. Its success was quintessentially American: because it was a phenomenon, its popularity itself was as much of a talking point as the movie, injecting a revived competitive spirit into Hollywood that hadn't been seen for decades. It redefined the parameters of a 'hit' – artistically, demographically and financially.

American literature was also in its pomp in a pumped-up period featuring Larry McMurtry, E. L. Doctorow, Toni Morrison, David Mamet, Hunter S. Thompson, Erica Jong, Saul Bellow, Robert Pirsig, Alex Haley, Thomas Pynchon, William Kotzwinkle and Tom Wolfe. Here, as elsewhere, cultural self-reflection was intense, feeding into what Wolfe would famously label the 'Me' decade – atomised individualism in a pampered world of self-development and psychological overindulgence.

The seventies became a decade of intense cultural self-reflection. And, as always, there were writers aplenty ready in the wings ready to throw their pennies into the conversation. A less self-absorbed project was Alex Haley's *Roots*, which told the semi-fictional story of Kunta Kinte, kidnapped from his African village to become a plantation slave in Virginia, then a soldier in the American civil war. Meticulously researched and viscerally worded, the publishers spent most of 1975 planning its launch. It won a Pulitzer Prize, remained at the top of the *New York Times* bestseller list for twenty-two weeks and, by 1977, had sold more than six million copies. Soon after, bookstores prepared for the phenomenon that became *The World According to Garp by* John Irving. It was the story of the life – and times – of T. S. Garp, the illegitimate son of a feminist leader who rapes a mentally infirm soldier to have a baby by herself. Dickensian in scope, the book charted Garp's life from cradle to grave, as a menagerie of characters – a former football star turned transgender activist, murderers, wrestlers, fortune tellers, cult members and a unicycling bear, to name a few – drifted in and out of his life. For many, it was their first exposure to an openly trans character and an openly asexual character in fiction, and the first book that confronted toxic masculinity and sexual violence head-on. It was a feminist novel written by a man and, in Irving's own words, 'an ode to the women's movement'.

By the time he published *The Philosophy of Andy Warhol (From A to B and Back Again)* in September, Warhol had already had more than his fair share of interpreters. The artist not only loved attention, he also enjoyed and actively encouraged misinterpretation. In this he was a lot like Bob Dylan, who was only too happy to let a generation of over-eager critics paint inaccurate portraits. Warhol had been called everything from an observer and pernicious corrupter to satiric reformer and charlatan. So, it was a surprise when he agreed to pen (with his frequent collaborator, Pat Hackett, and *Interview* magazine's Bob Colacello) an autobiography of sorts. a two-book deal and was also meant to deliver a biography of Paulette Goddard, which never happened. True to form, what Warhol produced only encouraged more speculation as to the veracity of his vision. The book was something of a hit, although it didn't tell you any more about Warhol than you already didn't know, it just presented the disinformation in a different way. There were sections of the book that told us about Andy's private life which were fresh and illuminating, but in sum it was yet another example of what one reviewer called his ability 'to make himself into a machinelike presence devoid of empathy'. At one point in the book, he says: 'If someone asks me, "What's your problem?" I'd have to say, "Skin."'

The continued popularity of British artist David Hockney was another example of just how in thrall to California we were in the UK. Hockney had emigrated in the sixties, and the lens through which he saw California became the way we all looked at it from the UK; it didn't matter how fractured or how nuanced Hockney's work was, his figurative rendering of bluer-than-blue swimming pools, brighter-than-bright skies and photographer Slim Aarons-style socialites looked impossibly glamorous. He was so famous; his work was used as the inspiration for the title sequence of the Herbert Ross film of Neil Simon's *California Suite.* Hockney had always excelled at capturing lifestyle, and in the seventies the California lifestyle was one that anyone could aspire to. They obviously couldn't have it, as California was a luxury that few in the UK could afford (most people couldn't even afford to visit), so by dint of that it became even more exotic.

California was at the forefront of sexual politics; San Francisco in particular had already become the most liberal city in North America. By 1975, the Castro district was established as the country's most prominent centre of what would become the LGBT community, a rainbow-coloured neighbourhood that was already a Mecca for young gay people all over the world. In 1975, California legalised consensual homosexuality, as Castro Street began to eclipse Polk Street as the cool gay address in San Francisco. Polk Street was a string of bars. The Castro was an entire district. The Castro had Victorian houses and churches, bookstores and restaurants, gyms, dry cleaners, supermarkets and even an elected member of the Board of Supervisors.

Homosexuality had been legal in the UK for eight years by 1975, but while most gay entertainers and rock stars were circumspect about their sexual proclivities, the general mood in the country was one of great tolerance. Especially in the gay discos of central London. A year later the city witnessed the arrival of Bang, London's first gay superclub. Held at the Sundown on Charing Cross Road every Monday and subsequently on Thursdays as the night's popularity grew, Bang could fit a thousand people, and had the loudest sound system in the West End.

The differences between American and British music were largely meteorological. In the UK, the cold and the grey were reflected in the edge and resilience of the songwriters and performers we produced; in the US, the music was as polished as the skies. Gentrification in Los Angeles promised a world that bore few similarities to the rock scene in the UK, which could often feel as prosaic and as underpowered as the economy. In LA there were palm trees, big mettlesome cars, cocktails, sunshine and a burgeoning, preordained luxury lifestyle. In the UK the trains were late, the food was awful and pubs always closed too early.

To us Brits, America was genuinely exotic, something we loved from a distance, but also something we would eventually betray. As record buyers, our interests were – with very few exceptions – mainly British and American. In the US, where the economy was stronger, music was less important and more commodified; in the UK, where cultural reverence was all, it was a period when we found it difficult to disguise our curiosity about American culture. It just seemed so much bigger, so much more fun and so much more glamorous. Of course, from a distance of fifty years, it's easy to forget that in the seventies, there was still some resistance to American culture in the UK, and even more so in mainland Europe. On one hand, there was a sense that US culture was reductive, like a hamburger: standardised, mass-produced and packaged. American network television was always used as an analogy: sponsored programming in which quality was sacrificed for reach and budget. On the other hand, there was international wariness surrounding the globalisation of US culture, a culture that was repeatedly accused of being slick, anodyne and neutered. 'Oh, it's so American,' was a common complaint, meaning the type of Americanism that equated size with success. Throughout the decade, America strobed like a lighthouse on all our horizons, whether we were happy about it or not.

Britain at the time was experiencing skyrocketing inflation and unemployment (stagflation, in other words), a wide range of strikes and states of emergency. The British therefore were suffering from a sense of despair and pessimism, scared the country was ungovernable. Consequently, life in Britain at the time could sometimes feel smaller and colder than in the US. The UK was still recovering from the three-day-week, which had been introduced by Edward Heath's Conservative government to conserve electricity, owing to industrial action by coal miners and railway workers. Television shut down at 10.30 p.m., pubs were closed and families ate dinner by candlelight. Football stadiums were banned from using floodlights, meaning matches had to be moved to weekday afternoons.

The measure was a major disaster for the Heath government, contributing to the losses in both 1974 elections, in February and October (both won by Labour's Harold Wilson). And yet the country still felt almost under siege, as Wilson's government attempted to stem Britain's economic collapse; so dire did things become that, in early 1975, five pence in every pound spent by the government was borrowed from abroad. It was understood that during the years of Wilson's leadership, his private and political secretary, Marcia Williams, was the single most influential figure among his staff. She exerted an unrivalled degree of power throughout his four periods as prime minister but became a source of conflict that clouded his years in Number 10 and would eventually diminish his political standing. She exerted a terrifying amount of control over Wilson, wielding lunch access to her boss like a weaponised press officer.

At one point, Wilson sneaked off from a dull function to do some work with his speechwriter, Joe Haines. Williams was furious. 'You little cunt,' she said to the prime minister, on finding him in his study. 'What do you think you're doing? You come back with me at once.' And so, he did. He always did. At the time there were rumours of a proposal to assassinate her, with Wilson's doctor suggesting to Haines and the policy chief, Bernard Donoughue, that they rid the PM of his troublesome mistress with a hefty dose of tranquillisers.

1975

Margaret Drabble's novel *The Ice Age* echoed the climate of malaise in a country 'sliding, sinking, shabby, dirty, lazy, inefficient, dangerous, in its death throes, worn out, clapped out, occasionally lashing out'. She readily admitted that 'the whole idea' for it 'came from reading newspapers'. She describes London as a sinking ship and the country as a whole in a collective state of decay and anxiety: prices are rising but the value of the pound is sinking; manufacturing firms are going into liquidation while other workers are striking. 'All over the country, people blamed other people for all the things that were going wrong – the trade unions, the present government, the miners, the car workers, the seamen, the Arabs, the Irish, their own husbands, their own wives, their own idle good-for-nothing offspring, comprehensive education.' As Drabble said, the ice age arrived when 'the flow had ceased to flow: the ball had stopped rolling'.

Drabble didn't invent the 'condition-of-England' novel, but as the *Paris Review* said, she did identify a gap in the market during the seventies, capturing the miasma of the age like a plume of smog under a bell jar. If there was one thing that separated both cultures it was sport (the British didn't celebrate baseball, basketball, hockey or American football) and the partisan nature of its supporters: and in the UK that meant football hooligans. By 1975, fighting at the football had become so commonplace that it had become an -ism, presented by officialdom as an ideology. It had become an accepted part of the experience. For instance, on 25 October, trouble flared at Upton Park, the home of West Ham, between their supporters and those of Manchester United. It was so bad that play was stopped for twenty minutes. Earlier in the year, Leeds United were banned from Europe after their fans rioted after the European Cup Final against Bayern Munich in Paris. This kind of behaviour was anathema to Americans and illustrated the differences in youth culture, which was as pertinent to music as it was to sport. Law enforcement, distance and patriotism also fed into the narrative, but fundamentally sports venues in the US weren't theatres of performative and tribal violence. Sport, like music, was largely entertainment.

Following Edward Heath's losses, Margaret Thatcher challenged him to a leadership contest which she won. 'For me it's like a dream,' said Thatcher, on being crowned the new Tory don in February 1975. The *Guardian* called her the new lioness of Conservatism, a queen-like figure in black taffeta – 'a pinkish tulip providing the only decoration'. How did she feel about facing the prime minister, Harold Wilson? 'About the same as he feels facing me. There is much to be done and I hope you will allow me time to do it thoughtfully and well. Yes, there are going to be some changes – a blend of continuity and change.' Even the people's paper found the need to ask her about the scepticism surrounding the appointment of a woman to the chief Tory job. 'Give me a chance,' she said.

On 5 June, Britain held its first referendum on whether the UK should stay in or leave the Common Market. The majority voted 'Yes' – 17,378,581 people (67.2 per cent) – to remain in Europe. 'Everyone should turn out in this referendum and vote "Yes", so that the question is over once and for all, we are really in Europe, and ready to go ahead,' said Thatcher in a TV interview. Labour was confusingly split over the issue. A suspension of cabinet responsibility allowed senior figures to campaign against the government position, with Tony Benn being widely mocked for encouraging a "No" vote. The UK press was largely positive, with only the *Spectator* and the *Daily Worker* aligning themselves with the Outs. With memories of the Second World War still relatively fresh, closer European cooperation was seen as a crucial means of avoiding future conflict. Suddenly, we had options. The twentieth century was turning into a century of choice. The Sex Discrimination Act came into force along with the Equal Pay Act on 29 December, aiming to end unequal pay between men and women in the workplace.

Don Revie managed the England football team (Alf Ramsey had been fired after England failed to qualify for the 1974 World Cup in West Germany), Richard Dunn had become the British Heavyweight boxing champion (he would unsuccessfully challenge Muhammad Ali for the World Heavyweight title in 1976) and James Hunt finished fourth in the Formula 1 championship (winning the following year). The most popular show on British television was *The Sweeney*, ITV's adrenalised detective show about the Flying Squad starring John Shaw as Detective Inspector Regan, with Dennis Waterman as Detective Sergeant George Carter. It was the first British police series to be shot on 16 mm film, entirely on location. Some said the star of the programme was the car they drove, a bronze three-litre Consul GT. Its success reflected the gritty reality of the decade. Here, the plain-clothes detectives acted no different from the villains. 'We're the Sweeney, son, and we haven't had any dinner – so unless you want a kicking . . .' A new album cost £2.99, a not inconsiderable sum, and an investment, almost. Even though albums sold in their hundreds of thousands, sometimes millions (in the UK, Pink Floyd's *Wish You Were Here* was the year's bestseller; in the US it was *Fleetwood Mac*), buying one wasn't a decision taken lightly. The radio hit of the year was 'Make Me Smile (Come Up and See Me)' by Steve Harley and Cockney Rebel, a glorious earworm that appeared to be pre-programmed into every pub jukebox. Originally a vengeful, gut-wrenching dirge, it had been overhauled by the producer Alan Parsons, who had ramped up the speed and turned it into a breezy feelgood jaunt that sounded like a love song. Apart from Harley's cockney lilt, it sounded as though it was built for car radios hurtling down Santa Monica Boulevard.

While there was a natural abandonment of sixties' idealism in Britain, many of the phenomena we often associate with the time – from denim, long hair and flares to provocative obscenity, the proliferation of class B drugs and ideological self-improvement (the Swinging Sixties affected less than one per cent of the population) – only achieved mass purchase in the seventies. Change didn't just affect the young; consequently, and complementing what was also happening in the US, the newly aspirational classes were satirised in Mike Leigh's *Abigail's Party,* Dick Clement and Ian La Frenais' *Whatever Happened to the Likely Lads?* and John Cleese and Connie Booth's *Fawlty Towers*.

If, by 1975, it appeared as though the seventies had finally arrived, change was imminent, especially in Britain. Even though almost all the progenitors of punk were American, it was the UK where it gained proper socio-economic and cultural traction because it was in the UK where it was most needed. In the US it was little more than a style, a style among many others which continued to blossom throughout the seventies. And so it was that the emergence of the Sex Pistols signalled the beginning of the end of a certain kind of American imperialism. And it started in 1975, a year that inadvertently celebrated artistry, ambition, and the refinement of a culture that had started twenty-one years earlier, way back in July 1954, when Elvis Presley recorded 'That's All Right'. If we can say that popular, post-war pop started that year, then it reached its very apex in 1975, a year which represented the very best of what a generation could achieve. The narrative arc had reached its crown. The seventies is still often painted as the decade that taste forgot (endearingly foolish fashions, embarrassing TV programmes, trite pop music, etc.), and yet for some the decade remains the high point of post-war pop culture. In some respects, 1975 is difficult to top.

1975 is an attempt to reclaim and redeem the mid-seventies, and to celebrate the ambition and artistic reach of the pre-punk culture, one that seemed to coalesce in the fantasy land of California. This period wasn't the cultural desert we have been led to believe by those who still treat 1976 as a cultural watershed. There were many great records released in 1975, yet the ones in this book are the ones that best realise their ambitions, the ones that best reflect the culture of the time – a culture of unrelenting depth, expertise and sophistication.

For recorded music, 1975 was one of the greatest years on record, a year that was responsible for some of the most mature and worldly records ever made.

IN 1975, I'VE CHOSEN TWENTY-ONE of the best records of the year, some of which were genuinely influential (Horses, '75, Another Green World), and others which were simply exemplary examples of high art

(The Hissing of Summer Lawns, Katy Lied). In each chapter I've taken a deep dive into the milieu which surrounded them, painting a picture of what these worlds were like: Californian cool, New York edge, London style, Jamaican imperialism, German technology, adult-orientated angst, ambient existentialism etc. For context, I've also included a Cultural Diary, a number of paramount cultural moments which impacted on the year, some of which had echoes of the sixties, and some which foretold the eighties in cinema, television, publishing, politics, terrorism, fashion, art, the theatre etc. Some are American, others not. As magazines were still so important to the culture, they've been included too (the NME, Rolling Stone, Interview, *the* Sunday Times Magazine, Street Life, Nova, Ritz, New West*).*

What if the Beatles had Never Existed?

The German equivalent of the Californian road trip, Kraftwerk's fourth studio album marked a step forward for the band; its novel combination of everyday noises with synthesiser explorations seemed both out of this world and firmly prosaic. The song itself would become their signature, even though it was a far looser interpretation of their computerised, orderly sound. What people don't realise is that Kraftwerk were designed to be the German Beatles, not as an act of homage, but a radical reaction to the Second World War.

Autobahn by Kraftwerk

ONE DAY IN MARCH 1974, Ralf Hütter and Florian Schneider were tinkering away in their studio in a fifties' building near the railway station in the centre of Düsseldorf. Today, like every day, they had arrived by pushbike, looking forward to furthering the development of German *Industrielle Volksmusik* ('industrial folk music'), specifically referencing a modern version of German regional musical traditions rather than *Doctor Who*-type music made in the UK.

Unlike Abbey Road or Sun Studios in Memphis, the duo's studio – Kling Klang – was deliberately anonymous. Situated in a nondescript street called Mintropstrasse, this yellowish building sat between a cheap hotel and a Turkish grocery store. A sex shop served as a reminder that it was close to the red-light district. It seemed nondescript for a reason – there was no reception, no phone, no signage, and nothing to tell you that inside was Germany's most important musical laboratory.

The week before they had bought a Minimoog synthesiser, which they were using alongside customised versions of the Farris Rhythm Unit 10 and various Vox Percussion drum machines.

They would typically spend eight to ten hours in the studio, although today they had a specific task, replicating the experience of driving along the *autobahn* (motorway) from Düsseldorf to Hamburg, including the industrial sounds of the Ruhr valley, the conveyor belts of the Bottrop and Castrop-Rauxel mining towns, and the rural Münster region, symbolised on the recording they were making by a (real) flute. They were also mischievously mucking about with other 'road' sounds, approximating engines, horns, tyres etc. 'Autobahn' would be finished at the studio owned by producer and musician Conny Plank in Cologne, but it was here, in this humdrum side street in the middle of Düsseldorf, where the future of European (and then American, and eventually global) music was forged.

The pair called Kling Klang their electronic garden, a space for them to work. They always thought of themselves as workers in sound rather than musicians or artists. They liked to say they were scientists rather than entertainers. They wanted to bridge the gap between music and technology. In their eyes, the studio was alive, an environment where man and machine might possibly become one. This was a command centre, with the musicians as quirky lab technicians twiddling knobs, static-like, shop window dummies. Even in 1975 it was full of kit, the floor covered in snaking cables. The basement of the studio was used to store old instruments and machines. The band never threw anything away and subsequently used the older equipment to accessorise the newer machines. The only windows were glass bricks.

Percussionist Wolfgang Flür described a typical day: 'In the Kling Klang studio of my time, we met up every evening around seven or eight. Then we would watch mostly TV news. After, we drank mostly coffee or went for an ice cream at a nearby ice cream shop. Then we went to the next room, which I called the rehearsal room – the "Kling Klang". And we made some Klang. Or Kling. It depended how we felt. Someone came up with a headline of a newspaper or maybe a TV report, then some melody was played around that theme. It developed over the following days, more and more. Lyrics came up, rhymes as well. And last, not least, a rhythm was drummed. That's how it worked.'

'I think we heard synthesiser sounds in the mid-sixties, around that time,' said Hütter. 'And through our cultural situation, radio from Cologne and electronic music – well, probably mid- to late-sixties, as students. We didn't know exactly which instruments were used, and then I think synthesisers came around at that time. I bought mine, I have to think . . . maybe '71, '72, around that time. [Synthesisers were] used for film music, and then we discovered that might be interesting for me. But I also worked the human voice and speech and language and poetry. It's kind of a concept of making my fingers sing.'

'That was the great release for me, when I saw . . . heard . . . listened to the Minimoog,' said Flür. 'It was just this fat sound of the ring modulators. If you take a guitar, it's always the same. But if you tune the volumes and the filters you can make millions of sounds out of [the Moog].'

Hütter and Schneider became obsessed with listening to the sound of driving, 'The wind, passing cars and lorries, the rain, every moment the sounds around you are changing, and the idea was to rebuild those sounds on the synth,' said Flür. When the proto-motorik beat arrives after about three minutes on the finished record, we are bombarded by a whole section where white noise has been modulated in a variety of creative ways – envelope retriggering, filtering, phasing, panning, and goodness knows what else. Unlike anything else on the market, unlike their previous records, and also unlike any other so-called 'krautrock' experimentations, Kraftwerk sounded genuinely different.

This had been their aim all along. As Joseph Beuys, Gerhard Richter and Anselm Kiefer had been doing in the world of conceptual art, and like Bernd and Hilla Becher, Thomas Ruff, Thomas Struth and Andreas Gursky had been doing in photography, Kraftwerk had been exploring new ways of recapturing a German cultural identity. Düsseldorf had always been a centre of the arts, but until the seventies its dominant art form was fine art.

1975

* * *

SINCE THE NINETEENTH CENTURY, DÜSSELDORF'S Kunstakademie had been renowned as one of Europe's most important art schools, and after the Second World War it became a leading forum for West Germany's re-emerging counterculture. The art school became a focal point for musicians as well as artists, and its large student body gave them a willing and captive audience. Kraftwerk were inextricably linked to the artistic community and had performed their first gig in 1971 at Creamcheese, the arts club that had been named after a Frank Zappa song, and which regularly held their own versions of Andy Warhol's Exploding Plastic Inevitable extravaganzas. The group wanted their music to be as different from the past as the new urban landscaping in Düsseldorf. Like many German cities, it had been reduced to rubble by Allied bombers. In the twenty-year period since the end of the war, the city had been rebuilt in a perfunctory modern style, an austere functionality that was ideological as well as born of necessity. German culture had been so discredited by the Nazis that architects didn't want to repair or restore the past. They wanted to sweep it all away and rebuild Germany anew, and consequently Düsseldorf was recreated as a New World city of bland apartment blocks, municipal buildings and highways. Kraftwerk had rejected the stereotypical Western orthodoxies of rhythm and blues and were challenging mainstream ideas of what actually constituted music. 'Germanness' wasn't exactly desirable immediately after the war, but as time went on, young German musicians began to find a way to express pride in their identity without embracing nationalism. And Kraftwerk were at the very front. The likes of Can, Tangerine Dream and Amon Düül had also explored this path, and yet their experimentation was not so far removed from the prog rock of their British cohorts. By the time of *Autobahn*, any avant garde notions had been binned. Kraftwerk wanted to be popular (even if it was on their own terms).

'My first ambition, musically, was to copy the Beatles,' said Flür. 'What else could I do? We had no music of our own and, in the sixties, there was no chance to play at any venues other than school halls, garden fairs, gymnasiums or private parties. The Beatles did something brilliant, but in the end, it was not new music, they came from rock and blues and they liked Elvis Presley and Chuck Berry and everything else. It's not a new invention in music I must say, even though they were amazing musicians, with their melodies and themes of course. But Kraftwerk did something more. We developed a completely new music, a new style. The press called it Krautrock, but it's not Krautrock really, it's electro pop.'

'I think it was more like an awakening in the late sixties of the whole living situation – the German word is *einfach Musik*,' said Hütter. 'Everyday music, like – it's more like discovering the tape recorder for us. Like, the world of sound: everyday life has a sound, and that's also why our studio is called Kling Klang studio because '*kling*' is the verb and '*klang*' is the noun for 'sound'. So it means 'sounding sound'. That's really what Kraftwerk is about. Sound sources are all around us, and we work with anything, from pocket calculators to computers, from voices, human voices, from machines, from body sounds to fantasy to synthetic sounds to speech from human voice to speech synthesis from anything, if possible. We don't want to limit ourselves to any specific sound like that was before when we were brought up in classical music. Then it had to be strings, it had to be piano, blah, blah, blah. We wanted to go beyond, to find a new silence and from there to progress to continue walking into the world of sound.'

Hütter and Scheider were born in 1946 and 1947 respectively, both of them into wealthy, upper-middle-class families in the staunchly Catholic Rhine-Ruhr region. The parents of both benefited enormously from Germany's post-war *Wirtschaftswunder*, or 'economic miracle'; Hütter's father was in textiles, Schneider's was an architect, and both sons had been brought up with a spirited sense of entitlement. This alone would give them the confidence to explore new ways of doing things.

In the years that followed the Second World War, Germany found itself in something of a cultural wilderness. With the country split between West and East, overt expressions of nationalism were frowned upon and anglophone music – at least in the West – began to fill the countercultural void. Hütter and Schneider were determined to find a new way to express German cultural identity without the impossible baggage of Nazism. They started by including German-language lyrics in their songwriting and ended up creating an entirely new form of music: futuristic, optimistic and extremely technologically progressive. They were obsessed with creating a new Germany, one that wasn't reliant on Western cultural influences. 'There's a whole generation in Germany, between [the ages of] thirty and fifty, that has lost its own identity, and that never even had any,' said Hütter at the time. 'There was really no German culture after the war. Everyone was rebuilding their homes and getting their little Volkswagens. In the clubs when we first started playing, you never heard a German record; you switched on the radio and all you heard was Anglo-American music; you went to the cinema and all the films were Italian and French. That's OK, but we needed our own cultural identity.' Kraftwerk's albums may have reinvented pop – Kraftwerk's future usually became pop's present – spawning half a dozen genres and helping define the digital age, but they were also perfectly in tune with the construction of contemporary Europe.

'Ralf had a kind of German idea in mind,' said Flür. 'Something with self-understanding and immaculate presence, after the ugly wars our parents had inflicted on the world.'

In 1974, as the rest of the music industry was reaching the kind of sophisticated peaks it had been seeking for two decades, German music was still in its infancy, scrabbling around for something fresh and clean and new. As FM radio in the US was creating a new platform for the type of music it took dozens of people and hundreds of studio hours to complete, extrapolating a rich and deep history of pop iconography, German music was striving for unanimity, trying to find its way in the world.

When Hütter and Schneider decided to feature German lyrics on their game-changing track 'Autobahn', it was a decision that amounted to a political gesture. It wasn't aggressive, though, and as their work developed, the idea of being German morphed into a broader European identity – as evinced by their 1977 album *Trans-Europe Express*. Identification with the political vision of a peaceful Europe was one way of dealing with the problem of being German in the aftermath of fascism. The creation of a specifically new German culture was also a way of putting distance not just between the past and the future, but also between Germany and the emerging neo-colonialism established by the continued presence of UK and US forces in Europe. This was especially true in Düsseldorf, as it served as an administrative base for the British Army of the Rhine (BAOR), which was stationed in the region, first as an occupying force, then as part of NATO's efforts to deter the Soviet Union from invading Germany during the Cold War. Inevitably, the long-term presence of British troops left its traces on the culture of the city along with the army's infrastructure. Local shops would cater for the troops, their families and their visitors from Britain; for example, bars would sell English beer and popular brands of blended whisky.

Hütter and Schneider wanted to build an autonomous type of music that would express a new German identity, one that rejected Nazism and Christian conservatism. Inevitably, it had to be constructed from scratch, in a laboratory, with machines. No guitars. No histrionics. No denim. Kraftwerk was born. Specifically, they wanted to anchor their vision of an authentic German music in their regional surroundings and traditions – 'ethnic music from the Rhine-Ruhr area', as Schneider put it.

'Country music, for example, is an impression of life which belongs to Texas,' said Hütter. 'This music, however, has nothing to do with Düsseldorf. We have always understood our music as specifically industrial music. And, therefore, also as ethnic music.' There was also a harking back to the past and, in that sense, Kraftwerk was a redemptive project, a reclamation of pre-war modernism and its aims. Kraftwerk rejected the nationalist mindset and would declare in interviews that they wanted to 'create a central European identity for ourselves', a multilingual, pan-European project. They were starting from scratch in a small town, a nowhereland, a cultural back room. The music had to be found. It was not there. For Hütter, the history of German music felt like archaeology. It felt old. Where was his sound?

Kraftwerk's first three albums, released between 1970 and 1973, were not that different from the other shaggy, experimental work of long-haired German hippies who were born just after the Second World War, but *Autobahn* was something else entirely. Being a transitional work, it still featured flute and even a guitar solo, but the beating heart of the title song was the Moog bassline and the lightly chugging electronic percussion, which, as they said themselves, 'hummed along in endless forward motion like glistening pistons in an air-cooled Volkswagen'. It was all about the beauty of repetition. On this record they were joined by Klaus Roeder and Wolfgang Flür and, even though they would have hated the notion (so old-fashioned!) they actually started to look like a band. Of course, their imagery was also based on something new. They didn't want to ape the generic look of most Anglo-rock bands, and so adopted the look developed by the artists Gilbert & George: Schneider and Hütter had seen them perform *The Singing Sculpture* (the pair sprayed with metallic coloured powders and singing the 1932 Flanagan & Allen song 'Underneath the Arches') in Düsseldorf in 1970. They were inspired by the idea of 'bringing art into everyday life'.

This was initially forbidding and Teutonic (like their look on the cover of *The Man-Machine*), but it soon became a kind of clubland blueprint, and during the late seventies would be copied relentlessly by the denizens of the Blitz club in London. The stylised covers of both *Trans-Europe Express* (1977) and *The Man-Machine* (1978) are retro fantasies playing into an old European tradition, but they were so arch that they simply looked playful. Your iconography could afford to be ironic when you sounded so new. Here was retro-futurism writ large. Kraftwerk wanted Kraftwerk to look and sound like no other German enterprise, like no other musical organisation, like nothing else defined as 'rock'. Not only did they achieve this, but they also laid the foundations for the EDM revolution of the twenty-first century. And as to the age-old question, were they as influential as the Beatles? Well, they were probably more so.

'We were brought up within the kind of classical Beethoven school of music,' said Hütter. 'We were aware there was a contemporary music scene, and of course a pop and rock scene. But where was our music? Finding our voice, I think that was the use of the tape recorder. Our contact to the tape recorder made us use synthetic voices, artificial personalities, all those robotic ideas. In the war, Germany was finished, everything wiped out physically and also mentally. We were nowhere. The only people we could relate to, we had to go back fifty years into the twenties. On the other hand, we were brought up in in the British sector and that's nothing we could relate to. There was no living musical thing other than the fifty-years-old musical thing or semi-academic electronic music, meaning psychologically we had to get ourselves going. And that has only been possible with our generation. You can see the generation before ours that is ten years older and they could not do it. The only thing they could do was get fat and drink. There was so much accumulated guilt that it physically took another generation to be productive, to be willing to say, "OK, I'm doing a song called 'Trans-Europe Express', or something . . ."'

Autobahn was released in the US in January 1975 and the world would never sound the same again. This year they would also pop up on British TV for the first time, appearing on the BBC's science and technology show *Tomorrow's World*. In the programme, the presenter Raymond Baxter said, 'The sounds are created in their studio in Düsseldorf, then reprogrammed and then recreated onstage with the minimum of fuss.' This appearance would later be heralded as the entire electronic ethos in one TV clip: 'The rejection of rock's fake spontaneity, the fastidious attention to detail, the Europhile slickness, the devotion to rhythm.' Baxter ended the report by stating emphatically that 'next year Kraftwerk hope to eliminate the keyboards altogether and build jackets with electronic lapels which can be played by touch'.

You certainly didn't hear records like this on the radio, not on daytime radio anyway. Like many who first heard it, I assumed they were singing 'Fun, fun, fun', in an homage to the Beach Boys, rather than 'Fahren, Fahren, Fahren'. But I liked the oddness of it, and the unlikely way it had become popular. I also thought it was probably a one-hit wonder. I definitely didn't think I'd be dancing to Kraftwerk in nightclubs in three years' time. I didn't think I'd be dressed like them either, in grey slacks and skinny ties? Don't be ridic! The idea of dressing like a 'straight' was still anathema in 1975, but then as soon as I saw the way in which Kraftwerk subverted normalcy, and how Talking Heads purposefully dressed like architecture students, things seemed different.

The German technocrats had suddenly electrified synth Britannia. At their best, Kraftwerk's music yielded a sensation of indeterminate depth and expansiveness, while appearing to do just the opposite ('It's good, but is it rock?' asked *Rolling Stone* when they were first exposed to 'Autobahn'). Their passivity was their strength, a refusal to wield guitars, express emotion or indeed to sweat. Minimal, robotic and exhilarating, they were a teasing glimpse of the future. If Kraftwerk counted as a science-fiction band, the fiction was just as plausible as the science. Visually, they were a conundrum: on the one hand they looked severe and authoritarian, and on the other, as lifeless as the stare of a shop-window mannequin.

Lester Bangs in *Creem* was also fascinated by Kraftwerk, the gonzo rock critic predicting that their music would herald a techno future at a time when many US critics were saying they represented the death of music. With his tongue firmly in his cheek, Bangs also ascribed the birth of rock and roll to Germany, solely because they had invented methamphetamine. Without which, he said, we would not have enjoyed the talents of Lenny Bruce, Bob Dylan, Lou Reed and the Velvet Underground, Neal Cassady or Jack Kerouac.

What Kraftwerk's fanbase were starting to understand was that their name meant 'power station': 'We play our machines into the electrical system and create transformed energy,' said Ralf Hütter. One of the group's nicknames was the Human Machine, another Sound Chemists. They were, and would remain, ruthlessly consistent.

We are the Enemy

1 January, King's Reach Tower, London SE1

IF THE INTELLECTUAL BUTTERFLY OR pop-cultural nomad had a bible in 1975 it was the New Musical Express, *which was probably the most important music title of the decade. The* NME *was how we discovered everything from esoteric pop music to galvanising politics and cynical OTT cultural criticism; it was our own national newspaper, a place where passion and cynicism were encouraged in equal measure. It was so important that at the start of the year they actually published a hardback book,* Greatest Hits: The Very Best of the New Musical Express.

The NME *was also the place you might learn about William Kotzwinkle (*Doctor Rat*), Robert M. Pirsig (*Zen and the Art of Motorcycle Maintenance*) and of course Hunter S. Thompson and Tom Wolfe. It was a commercial title fed by the independent sector, a mainstream paper staffed with outliers from the underground press. And it was all because of Nick Logan, an ex-mod from northeast London. Logan joined IPC's* NME *as one of five staff writers in 1967, when the weekly was still reliant on the major stars of the earlier part of the decade such as the Beatles and the Rolling Stones. As rock supplanted pop and album sales overtook those of singles, the paper lost ground to its more austere, momentarily highbrow rival,* Melody Maker, *which was also published by IPC.*

In 1972, IPC management promoted Logan's colleague Alan Smith to editor, with assistant editor Logan given the brief of overhauling content and design. Together, they assembled a new editorial team drawn from the ranks of the underground and independently published press, including Charles Shaar Murray from Oz, *Mick Farren from* International Times *and Nick Kent and Pennie Smith from* Frendz. *Tony Tyler also joined, as well as Ian MacDonald, who would go on to write a seminal text on the Beatles,* Revolution in the Head. *A year later, at the age of twenty-six, Logan assumed the editorship from Smith, becoming the youngest national newspaper editor in the UK. During his tenure, which would last until 1978, the paper regularly sold in excess of 250,000 copies a week. As future contributor John May would say, this was New Journalism of an irreverent, drug-fuelled kind that captured the spirit of the times, inspired by the likes of Lester Bangs and* Creem *magazine. There were music, books, film, youth culture and politics.*

'I had some cause for self-congratulation,' said Nick Kent, 'for I was now strapped mind, body and soul to the whirling zeitgeist of cutting-edge popular culture until I could feel the aftershocks puncturing my very bones.'

Like Hunter S. Thompson, Nik Cohn and Lester Bangs, there existed around Kent an almost mythic glow. Degenerate poseur, celebrity drug addict and genius wordsmith, he was a man who has lived rock and roll to the full. During the seventies, Kent was as famous for his drug intake as he was for his journalism, and from 1974 was addicted to heroin, cocaine, methadone and various tranquilisers. 'At the age of nineteen I started smoking hashish in earnest,' he says, 'and then moved on to speed and cocaine. Finally, I was offered some heroin, and that was it for me. It was like being in heaven, it was ecstasy. I thought this is worth getting lost for. But I got hooked and it took me years to kick it.'

With a knack for decoding those stars who displayed little but a desire to be famous, and who rebelled against nothing but personal imposition, Kent himself rarely hid his light under a bushel. After all, he was a star. One evening, Kent was dining with David Bowie and Iggy Pop in a little Chinese restaurant in Gerrard Street, in London's Soho. Confronted with a triumvirate of pop icons, the confused, if star-struck, waitress approached the man who most looked like a rock star for his autograph. Bowie and Iggy were shocked. Nick was flattered.

Charles Shaar Murray (known as CSM) was almost as famous as the acts he wrote about. 'At the NME *we weren't king makers, but for a brief period I was a conduit,' he says. 'David was always very good at picking people up when he wanted something from them. In a genial way he was quite ruthless and a user. I started to suspect this when people started to get a bit snarky, saying that I was his representative on Earth. One thing I became aware of quite quickly was that a lot of journalists he'd dealt with in earlier stages of his career, some of whom had actually been very supportive of him, were suddenly not flavour-of-the-month and he wouldn't talk to them any more. For a certain period, I was the person he spoke to.*

'An example of this is his [Ziggy Stardust] *retirement concert at Hammersmith Odeon. The week of the Hammersmith concerts, I had a week's holiday booked in Cornwall with my then-girlfriend. I'd seen several shows on that tour, so I thought, I know it's the end of the tour, but it's just another show. I've seen it. I've got this holiday booked, I'm a bit knackered and I've seen a lot of his gigs. A few days beforehand, Jeff Beck's girlfriend had rung me up and said that Jeff would quite like to go to this show, and I'd called Bowie's office and asked them to sort it. They said "Of course," because he was Mick Ronson's idol. Then when I was on the phone, they said, "Hold on, David wants to talk to you." He came on the phone and said, "Look, I'm knocking it on the head after the final show. Don't tell anyone apart from the people in the office because even the band don't know yet." He said he wanted the story on the front of the* NME *the day after the final gig. He knew when we went to press and, as the gig was on the Tuesday, there was just enough time to do it. The paper was on the streets in London on the Wednesday, and on the Thursday in the rest of the country. I went and told Nick Logan, I gave him the quotes and we put the front page together, and the front page was actually coming off the presses before Bowie had actually announced on stage that he was killing off Ziggy Stardust.'*

In 1972 I was still buying Popswap *each week, a kind of proto-*Smash Hits *that was full of pictures and interviews with glam stars like Bowie, Gary Glitter, Marc Bolan and Bryan Ferry. It was full colour, glamorous, and perfectly pitched at twelve-year-olds who were taking more interest in the pop charts than their schoolwork. Then one day in February '73 (as I had just turned teenage) I saw a copy of the* NME, *and more importantly a Nick Kent review of Alice Cooper's* Billion Dollar Babies LP. *I loved Alice Cooper (I still have massive scrapbook I spent months compiling that winter) but had never heard of this writer. At the time I'm not sure I had heard of any writers, and couldn't believe how caustic he was, how funny, and how dismissive. It started like this: 'You've got to hand it to Alice Cooper and the boys – they know just when to pump out another album for the kids to tether their fantasies on', and you can still find it on* Rock's Back Pages. *From then on, Alice Cooper became a much less important part of my life, as Nick Kent practically moved in.*

By 1975 I was a devout consumer of the NME, *and – like the other 250,000 people who read it (not that I knew that, obviously) – considered myself to be the king of snark.*

London, 1975: Cocktails, Crooner Pastiche and Deco Retro

> If you were a certain kind of London person, the only place to meet and drink in 1975 was a member's club in Great Queen Street in fast-smartening Covent Garden, the home of accountants, low-level lawyers and – crucially – recently arrived ad agencies. The club was called Zanzibar, run by a wine merchant and an architect, and anyone who was anyone went there. It had a long, wavy, mirrored bar, which meant if you sat on one of the tall, chrome barstools, you could talk to everyone at once.
>
> *Siren* by Roxy Music

THE ZANZIBAR WAS A VERY social place. It was deco-ish, served cocktails, and was just about the most modern place in town. As Peter York, Zanzibar member and all-round style guru said, 'The cocktail thing, deliberately lurid and sticky, was self-consciously period, like Zanzibar's cream walls and the long, curved, black-lacquer bar. They'd been looking at old photographs. Zanzibar introduced media twenty-somethings on the make to the whole cocktail ritual, the icing and slicing and shaking.' It was an art-directed existence.

There were other places to convene in London – Langan's Brasserie, the Poule-au-Pot on Pimlico Green, the Soho Trat in Romilly Street, and Parsons, on the bit of the Fulham Road people now call the Beach – but nowhere nearly as zeitgeist defining as Zanzibar. There were so few fashionable people in London, and this is where they all came to eat and convene. This was also the year that saw the final days of *Nova* magazine, the last hurrah of Biba, and hordes of girls in diaphanous Daisy Buchanan dresses and henna bobs. The year that Derek Jarman won the Alternative Miss World, the big Them event staged at the Butler's Wharf Thameside warehouse that housed Jarman and Andrew Logan, the pageant's creator and master of ceremonies. It was also the year Fiorucci opened its doors on the Kings Road.

These parts of London obviously had a soundtrack, one that was driven by gay soul, Bette Midler, Manhattan Transfer (the 'Man-Trans'), Cockney Rebel, the *Cabaret* soundtrack and Liza Minnelli's *Liza with a Z*, and the twin delights of David Bowie and Roxy Music. Both the latter had managed to co-opt the fringes of a generation who weren't content with the orthodoxies of the rock milieu and would be cited as harbingers of change for the rest of the decade in one way or another. In 1975, they adopted the doctrines of the disco, Bowie (who at the time tended to be slightly ahead of the curve) in February with his single 'Young Americans', and Ferry's Roxy Music with 'Love is the Drug', from the *Siren* album, in December (Nile Rodgers says that John Gustafson's bass line was a big influence on Chic's 'Good Times'). Bowie and Ferry were the lodestars of '75: the *Siren* tour featured Ferry and his two glamorous backing singers costumed in khaki GI chic, inspired by Elvis Presley and created by the fashion designer Antony Price; Bowie, meanwhile, adopted a henna'd wedge and plastic sandals. Both were mercilessly copied. 'Fans dressed for the concerts with meticulous attention to detail,' says the writer Michael Bracewell, 'creating a total look that mirrored the super-stylised spectrum of imagery that Roxy Music merged into a single artistic statement.' Their devotees were so attentive, so analytical in their dress, that in 1973 the *New Musical Express* ran a review of the audience at a Roxy gig, rather than the concert itself.

'You'd go to see Roxy Music and everyone would be dressed as Bryan, which was absolutely fascinating,' said Peter York. 'You could see that thing of people following as closely as they could their great heroes, and there was already that thing that Robert Elms described later as pale faces and white socks, which is a combination of art students and aspiring working-class kids of all kinds, from all sorts of regions. Fabulous audiences. My eyes were out on stalks for the audiences.'

Bryan Ferry said, 'Roxy were always about pop art, about accessibility, about the street and, contrary to popular opinion, we were not elitist. Pop art was about celebrating mass culture. We were also quite a lads' band back then and we appealed to the boys on the street. It's when we reformed that we became a ladies' band. Dressing well is not exactly rocket science. And in my case, it's always been secondary. I have, however, always been a keen student. Antony was the real architect of the Roxy look, and he made these amazing, sculpted clothes. Whether he's designing for men or for women, he really understands bodies, understands what looks good, what looks sexy. When I was younger, I was a mod, and I always appreciated the high street – every mod did. Mod was really all about dressing to a certain level without having to spend a fortune. It was about great design for everyone and being smart.'

The consensus says that every Roxy album was half as good as the one before it (with fans debating whether *Roxy Music* was really better than *For Your Pleasure)*, but that is a nonsense. In fact, the final proper Roxy albums – *Stranded*, *Country Life* and *Siren* – might just be the very best things they did. By the time of *Country Life*, Ferry was already exploring America – 'Prairie Rose' being a rampaging ode to Texas. Roxy's success lies in their calculation, and then their delivery. The music is at times overwhelming, as Roxy made the kind of noise that swirled around you, forcing you to engage with it even as it pushed you away. Roxy Music: cold and warm at the same time, like the breeze.

Speak to any Roxy aficionado about their best album, and the conversation will focus exclusively on whether the answer is Roxy 1 (their debut) or Roxy 2 (*For Your Pleasure*). I've always maintained that the tussle should be between Roxy 3 (*Stranded*) and Roxy 4 (*Country Life*) but, after decades of arguments, I still don't have a convincing answer. Both 3 and 4 are less eccentric than the first two (which were made with Brian Eno), and there is an orthodoxy about them which critics slightly sneer at, but there is also a slickness and a confidence that is still bewilderingly seductive. Those same critics have been unnecessarily brutal in their appraisal of *Siren*, although I'm sure their initial dissatisfaction stems from the not unrelated facts that Bryan Ferry had started to embrace disco (with 'Love is the Drug' and 'Both Ends Burning') and was going out with model Jerry Hall. Some band members also felt the record was below par, seeing that some of them were also pursuing solo careers, and so mindful of publishing credits that they got even more involved in the songwriting. Nevertheless, the producer Chris Thomas pushed the band to strive for perfection, which is why the record has lasted so well. Ferry was still pouring himself into Roxy; the band tended to work on finishing their songs instrumentally, with Ferry leaving the lyrics (and therefore the tune) to the very last; with 'Love is the Drug', he came up with an almost complete vocal line with fantastic lyrics to general amazement and applause in AIR Studio No. 1. Lyrically, 'Just Another High', which ends the album, was also blindingly acute, and is justly regarded as one of Ferry's best.

For me and my sartorially-obsessed friends, 'Love is the Drug' and its disco cousin 'Both Ends Burning' bridged something of a gap, saliently being Roxy Music records you'd hear in a club rather than just on our turntables. I wasn't sure *Siren* was as good as *Country Life,* but there were days when I did. What was never in doubt was Ferry's ability to disregard the rock-and-roll rat race. Like David Bowie, he invented personas; unlike Bowie, he became haunted by them. My friend Fiona (Dealey, who would soon become a formative Blitz Kid) had been to see Ferry play a solo gig at the Albert Hall the year before. 'I wanted the whole thing,' she said. 'I loved the way he dressed and was completely in love with him. Then I looked up at the VIP area, and I could see a little group of people who looked amazing, had the best clothes on... One girl had a pink parasol perm, and I kept looking up at them and I thought, That's the cool gang. That's what drove me.'

It's what drove so many of us.

Some thought Bryan Ferry woke up every morning and put on a tuxedo. But in reality, he always waited until the cocktail hours. With 'In Every Dream Home a Heartache', from Roxy's second album, *For Your Pleasure*, Ferry established himself as the decade's quintessential boulevardier, a man of wealth and taste. With *Siren*, they would reach their apotheosis, finessing their languorous urban ennui, discovering the disco, and fusing all their tricks and treats.

Ferry was the Tyneside proto-lounge lizard who, along with Bowie and Marc Bolan, helped invent the feel of the early seventies. Mixing the glamour of nostalgia with edgy modernity is now nothing special, but Ferry experimented with this long before most. When Roxy Music first appeared, at the very start of the seventies, they were so different from their peers that they may as well have been from outer space. Roxy's music was certainly arresting – pop-art Americana mixed with searing R&B and avant-garde electronics – but it was their clothes that really turned heads. Leopard and snakeskin. Gold lamé. Pastel-pink leather. At a time when most pop groups thought long hair, cowboy boots and denim were the height of sartorial elegance, Ferry and his band walked into the world looking like renegade spacemen. The first Roxy Music album was a whirlwind of playful decadence, a members-only nightclub with Ferry as the quintessential playboy making a bumblebee exit, buzzing from table to table on his way out.

Roxy were a montage of hot music, giddy, drunken laughter, stoic, almost minatory posing and nightclub reverie. The music took the old sweet rock and roll melodies and twisted them like hairpins. This was a seemingly classless world full of bright lights and dark alleys, of loose women wearing tight dresses and bad men wearing good suits: a glamorous world etched with danger, a full-scale escape from reality, Roxy were its fulcrum, the very zenith of Style Centrale.

Much like the generation he inspired, as a boy Ferry had longed for escape. Having grown up in the north-east of England in a fairly impoverished mining village called Washington, near Durham, Ferry was desperate to travel south, to London, to metropolitan excess. He was helped in his cause by the legendary pop artist Richard

Hamilton, who had taught him at Newcastle University. Hamilton not only inspired the Roxy Music cover-girl album images, he also inspired the music itself, a mixture of sci-fi aspiration and fifties' Americana. 'To go that route seemed the only option,' says Ferry. 'I mean, we could've had a picture of a band, looking rather glum, which was normal, standing on a cobbled street or something. But I didn't fancy that. The pin-up was a great way to sell things traditionally, whether it was a Cadillac or a Coke bottle or a packet of cigarettes.'

Or even Roxy Music, come to that.

In 1957, Hamilton had come up with a definitive description of pop art, one that could easily apply to Roxy itself, and one that Ferry took to heart: 'Popular (designed for a mass audience), transient (short-term solution), expendable (easily forgotten), low cost, mass produced, young (aimed at youth), sexy, gimmicky, glamorous, big business.' Throw in some fifties' balladry, white noise and synthesiser treatments, plus some futuristic teddy boys and there you had Ferry's vision.

Even their name was pop: 'We made a list of about twenty names for groups,' says Ferry. 'We thought it should be magical or mystical but not mean anything, like cinema names, Locarno, Gaumont, Rialto . . .' According to Andy Mackay, Roxy's bequiffed sax player, 'If Roxy Music had been like cooking, it would be like a dish in Marinetti's futurist cookbook, *Car Crash*: a hemisphere of puréed dates and a hemisphere of puréed anchovies, which are then stuck together in a ball and served in a pool of raspberry juice. I mean, it's virtually inedible, but it can be done.' A fashion grenade rather than a fashion parade (Roxy were roughly analogous to a Gucci or a Chanel, only with a yard of fake leopard skin and a sachet of sequins thrown in for good measure), Roxy transmitted aspiration, travel, power . . . sex. It was a modern-day love letter to an anglicised America, a seventies' idea of fifties' space nobility. They were conquering the new frontier. In silver jump suits! Playing guitars! In front of girls!

'Bryan was wonderful – very clever, agonisingly fastidious, he managed to give the postmodern appearance of standing outside what he did – then a new posture – while at the same time, being hyper-involved and passionate,' said Peter York. 'He was ironical *and* sincere. When you heard "Virginia Plain" for the first time, you knew you were listening to the natural antagonist of Joe Cocker; you were listening to a singer whose whole approach said, "I'm not *singing*, I'm *being a singer*." Similarly, *the looks*, all of them said, this is just a phase, just a costume for now. But at the same time, it mattered enormously how you looked. So you got dyed feather shoulders over blue spangles, or you got the white tux with half-tamed forelock, or you got the Antony Price GI combination of 1975 (with matching girl back-up singers, very sexy) and, whatever you got, it was done beautifully. And thousands of boys and girls who were sick of the street, sick of the High Street, low-life, no-life, did a Bryan and turned up at the concerts wearing the Bryan look (or the gorgeous girl accessory-to-Bryan look, as in those first wet-dream Roxy album sleeves). People went to the first concerts at the Rainbow, in Finsbury Park, and had their lives changed.'

Although they didn't know it at the time, Ferryphiles were part of the first postmodern generation, while Roxy Music were the first postmodern pop group. Ferry courted the art-school crowd and paid lip service to the fashion industry while never forgetting the ossified denim-clad punters who paid his wages. Essentially, Ferry discovered his own identity via the assumption of false ones, in a bid, perhaps, to spend the rest of his life in a world full of Roxy Music album covers. Using a little bit of Billy Fury, a little bit of Biba and a soupçon of good old-fashioned music hall, Ferry reinvented himself as a glittering, larger-than-need-be playboy. And boy did he love his job. He was brazen in his lust for a life that was never really meant to be his. He wanted not just the fast cars and the fast women, but the kudos that came with them. It wasn't enough for him to be seen to be enjoying himself, people had to believe that he meant it, that these things for him were not merely confetti – they were food and drink.

Occasionally ridiculed by the press, he was adored by the faithful. So cool was he when Roxy were in their first flush of success – 'Virginia Plain', 'Pyjamarama', 'Do the Strand' – that he became, almost overnight, a unique arbiter of style, so influential that fans would copy every little detail of his dress, whether he was trussed up like a fifties' retro-future crooner, an American GI or a tuxedo-clad lounge lizard. Marco Pirroni, who would become the guitarist in Adam and the Ants, was such a Ferry devotee that he once used a magnifying glass to study a photo of Ferry just to see what cigarettes he was smoking. 'Even the cigarettes were really good, because they were all white [St Moritz]. It was an important statement,' Pirroni says.

Ferry's own style icon was the fashion designer Antony Price, who created many of Ferry's best looks, and is someone who deserves proper recognition. 'Antony Price reinvented the suit so that it was no longer about going to the office,' says the stylist David Thomas. 'He made it rock and roll. He started at a time when British fashion didn't have sponsors. It was the era before the superstar designer. They all came after him. Yet he was a visionary. He created that military, dandy, sexy, eclectic men's look. He created rock and roll fashion.'

Ferry's fastidiousness played against him – critics mistook his fondness for archness for phoniness. He was, however, revered by many of those who instigated punk, even though they were reluctant to say so at the time. 'I always do sound tests using Roxy Music, chunks of songs [notably "2 H.B."],' says the Sex Pistols' John Lydon. 'Bryan Ferry? I mean, he's not the most talented singer, but I love everything he's ever done.' (In 1976 Martin Amis would review the Rolling Stones at Earls Court for the *New Statesman*. 'The ante-hall of the Earls Court Arena was a Brobdingnagian underground car park of remote and overcrowded bars, sweet shops and dirty hot-drinks machines. Normally a token homogeneity obtains at the average rock concert: David Bowie fans all look and behave like David Bowie, Bryan Ferry fans all look and behave like Bryan Ferry etc. But everyone is a Stones fan.')

Ferry had an ability to try a little harder than many of his peers, even if he didn't always convince. The *NME*'s revered critic Nick Kent eloquently dubbed him 'the George Lazenby of the Argentinian corned-beef market'. However, most people thought that Ferry was so smooth, Gucci wore his shoes.

'I met Bryan in 1972 at a party in London,' said Antony Price. 'He came up to me and asked if I wanted to get involved with Roxy Music. So, a few days later he came round to mine with his picture book, and he pulls out the record and I put my headphones on and I hear the first Roxy album. I loved his voice and I could see he was a nice-looking guy and I thought he was going to look good, and that's how it kicked off. Back in the sixties, Stax put a woman on the cover of *Otis Blue*, the Otis Redding album, otherwise it wouldn't have sold to a white audience; but Bryan liked the idea of using a woman to sell a product. For the first Roxy album he wanted this Rita Hayworth look, and [the model] Kari-Ann was the closest you could get to Rita Hayworth, as she had this stunning red hair. Bryan also understood that women don't care whether a man is gay or straight, they won't stop fancying them, in fact it doubles the fun because it means if you can pull him in then you've really got it. Roxy didn't want to be seen as gay, but they wanted to be seen as exotic. Bryan didn't really care either way, as he knew it was all about how you looked.'

The clothes Roxy were wearing at that time would have put off quite a large chunk of people. What Ferry liked about the American bands, the Stax label and Motown, they were into presentation and show business, mohair suits, quite slick. And regarding the cover art, he thought of all the American pop culture icons – Marilyn Monroe, Elvis – and selling cigarettes or beer with a glamorous image. But it was a bit off-kilter as well; there was something a bit strange about it, futuristic as well as retro. When the press talk about the early seventies, they always call it glam rock, but Ferry and Price didn't call it that. They just thought it was good taste, dressing smartly and living well. They both had a sense of perfection, and *Siren* was probably where it all came together best.

'If Roxy were the very essence of retro-futurism when they began, embodying the glorious confusion wrought by the possibilities of the space age colliding with the solidly nostalgic sensibilities of the fifties, by 1975 their febrile alchemy had settled somewhat,' said the journalist Graeme Thomson. 'On *Siren* the sound pushes forward with a relatively solid sense of purpose, rather than shuttling back and forth with mad abandon. The band are cruising into the great beyond with a streamlined purr, well past the halfway point on the road between "Virginia Plain" and *Avalon*. Some critics asked, where is the Art? – a grumble that still surfaces periodically – but it's there, all right. You'll find it in the occult guitar, plink-plonk atmospherics and wandering oboe of the first two minutes of "Sentimental Fool", a song that harks back to the magnificent strangeness of *For Your Pleasure*. It's here, too, in "She Sells", a delightful tumble of a track, which begins like the Doobie Brothers fronted by David Niven, then takes a series of lurching handbrake turns into furious urban funk and robotic art rock; and in Andy Mackay's ululating sax solo on "Nightingale", which otherwise is one of *Siren*'s less arresting songs, despite its prettily folksy intro and Paul Thompson's driving drums.

Most obviously, it's there in Graham Hughes's striking cover image. By the time of *Siren*, Antony Price had been working with Bryan for three years and they had done four albums together – *Roxy Music*, *For Your Pleasure*, *Stranded* and *Country Life* – so they were well adapted to working together. Ferry always used to call Price the batsman – being a Libra he would bat things back at him, building on his ideas. Ferry and Price drew sketches and then Hughes took the pictures. Ferry had written a song called "Siren", and for the cover he decided they should have a mermaid. Ferry's soon-to-be girlfriend, Jerry Hall, is presented as a dangerously alluring vision in aquamarine – love very literally on the rocks – near Holyhead lighthouse, South Stack on Anglesey.'

'The cover of the new Roxy Music album is credited to eight people, two more than made the music,' noted *Rolling Stone*, somewhat archly, in their review. There was something reassuring about that. Roxy were changing, but they were doing so with their sense of style intact.

The idea for the location was Ferry's, after he saw a TV documentary about lava flows and rock formations in Anglesey, in which South Stack was heavily featured. On the shoot, Ferry held an umbrella over

Hall as it started to rain. She at the time was living in Paris with Grace Jones and the fashion illustrator Tony Viramontes, and this location couldn't have been more of a culture shock. The shoot took place on the hottest day of the year, the sea was completely blue, and as the costume was green Price had to change it to blue by spraying it.

'Bryan was at the height of his fame when I arrived in London ahead of the shoot in summer 1975,' says Jerry Hall. 'I loved Roxy Music and thought Bryan had the most beautiful voice, heartbreakingly touching and sexy. One look at his elegant, handsome face and I forgot all about New York. On my first night in London, he took me out to dinner in a black Jaguar with leather seats. When he shifted gears, his hand almost brushed my knees – there was a lot of chemistry between us. The album photoshoot was in Wales, where we stayed in a little seaside hotel. After dinner, I went to bed and curled my hair for the following day's work. I was tucked up in bed in my nightdress when Bryan knocked on my door. I let him in and got back under the covers, embarrassed.'

Ferry had managed to metamorphosise. Where once he'd been a boy on the outside of the glass, he was now tightly inside, and there was no way he was going to risk slipping out again. He'd got the look, he'd got his gang, and he'd got his girl. He'd also got his public, a public who didn't awfully mind that he wasn't all he appeared. After all, neither were they. Essentially, he had been smart enough to understand he could make a career out of pretending to be someone who pretended. All he needed was the talent. David Bowie once said that during the seventies, he and Ferry were both independently keen to let people know how smart they were, 'that we weren't all trying to be Chuck Berry. I knew Ferry was a huge Dada fan, for instance. He even did an album called *The Bride Stripped Bare* [after the artwork by Dadist Marcel Duchamp]. Eno and I went, "He shouldn't do that," thinking we should have done it first.'

Martin Fry would channel Roxy's visual messaging with his group ABC in the early eighties. 'On the *Siren* tour, everybody showed up in their tuxedos, which he'd worn on the previous tour, but Bryan Ferry came out in his GI outfit, with Jerry Hall, which was an extraordinary sight – *completely* glamorous.'

'The *Siren* tour was an extension of the military look, and I turned Bryan into a GI,' said Antony Price. 'The air hostess dresses that the back-up singers wore were so tight and so sculpted they had to be lifted onto the stage as they couldn't move their legs. The army look was happening at the time, as everything had to look very American. I quite liked it, and I was slightly military-orientated with my design anyway so that crept into it and Bryan looked good in it. I know a lot of people in the clubs started wearing it, and I actually went down to suss out the Goldmine in Canvey Island to see if we could do a shoot there. We didn't. They were doing a forties night and had swing music. Obviously, a lot of people would copy what Bryan wore. At their concerts I would peak through the curtains and there would be people in the front row wearing copies of the clothes, so I was aware the clothes had become very important. They were what Roxy were all about.'

The Iron Lady

11 February, House of Commons, London SW1

WHEN MARGARET THATCHER TOOK CHARGE of the Conservative Party in February she was still being treated as a novelty, a woman who had the audacity to do a man's job. As shadow environment secretary she had been angered by what she saw as Edward Heath's U-turns on Conservative economic policy and so she stood against him.

When she went into his office to tell him of her decision, he didn't even bother to look up. 'You'll lose,' he said. 'Good day to you.'

Everyone else thought she'd lose, too, including all the Conservative papers (the only publication to call the result was the Spectator.*) The political establishment in 1975 was, to an extent that is hard to imagine now, almost exclusively male. Schools (at least the schools at which Tory ministers were educated) were single sex and the political clubs in which plots were hatched were all male. Thatcher also had no experience of foreign and defence policy (crucial areas for a prime minister). The latter was particularly important because Britain in the seventies was still governed by the generation that had been formed by the Second World War. But in the end, none of this mattered.*

At a news conference after her win, which she ran with her characteristic mixture of smiling politeness and total control, she fended off questions about the principal political theme of the day: how much did she intend to shake up the party? 'There must be a blend of continuity and change,' she told a reporter, giving no indication just how dramatic her strategies were going to be, or how brutally they would be delivered.

53

She didn't look like a giant killer – not that Heath was feared much by many in his party. In her twinset and pearls, perfectly coiffed hair, practical handbags and civic sense of style, the forty-nine-year-old seemed almost demure. She favoured tailored, cobalt-blue skirt suits, pussycat-bow blouses and sensible shoes. Then, of course, there were the voice coaching lessons, during which she learned to lower her pitch and talk in slow, strangulated received pronunciation English. Apparently, she did this artificially, and without informed advice, which made her speaking voice sound even stranger. Impressionists were told that if they wanted to successfully imitate her then they needed to try speaking on the edge of a breathy yawn. Thatcher was a great admirer of the Queen (the admiration travelled in one direction) and went out of her way to try and emulate her way of speaking. She was attempting to make herself sound warmer, and less strident, whereas she ended up sounding like a Doctor Who *baddie as imagined by Barry Humphries's Dame Edna.*

However, she had already developed an icy stare, which she deployed to skewer anyone who had the temerity or the witlessness to challenge her unswerving certainties. Her gaze would swivel on reporters like a tank-barrel. Which meant she was good copy. Conflict always makes for headlines, and when the media realised she was serious about her ambitions to change the country (starting with her own party), they started to take notice. With Thatcher, we would soon learn, conflict was never in short supply.

The first woman to become prime minister and the first to lead a major Western power in modern times, she was hard-driving and hard-headed, while the pungent economic medicine she administered to a country sickened by inflation, budget deficits and industrial unrest brought her wide swings in popularity, not least in the media. In the Commons her performances, first against Harold Wilson and then James Callaghan, were initially disappointing: she was shrill, over-briefed and often lacking in confidence. But she soon learned to trust her convictions, making mincemeat of anyone who disagreed with her. Although she would soon dominate her party as well as the government machine, her self-image was of an outsider battling with an inert establishment.

She didn't like the establishment and didn't care much for Government; evening visitors to the flat above Downing Street would often find her and her husband, Denis, watching the news and grumbling about the state of the nation.

For a woman so intent on root-and-branch reform, who was obsessed about creating a new country based on principles buried in the past, when she arrived she looked incongruously old-fashioned, like a caricature from another generation. She would change Britain immeasurably, and yet she always looked as though she was doing it in character.

At the time she became leader, she had not yet fully embraced the principles and policies which were subsequently developed into Thatcherism. She instinctively perceived that the path chosen by the post-war Conservative Party had been a mistake, and her experiences as a minister in Heath's cabinet reinforced her growing conviction that the Tories needed a change of ideological direction. Not even party diehards could have imagined what she was now going to do. Under her leadership, the Conservatives would shift further to the right, calling for privatisation of national industries and utilities and promising a resolute defence of Britain's interests abroad. When it was announced she had won, the MP for Finchley, north London, since 1959 rejected suggestions of great celebrations. She said, 'Good heaven, no. There's far too much work to be done.'

She would end her first party conference speech as leader in October with the following: 'We are coming, I think, to yet another turning point in our long history. We can go on as we have been going and continue down. Or we can stop – and with a decisive act of will we can say, "Enough." Let us, all of us, here today and others, far beyond this hall who believe in our cause make that act of will. Let us proclaim our faith in a new and better future for our party and our people. Let us resolve to heal the wounds of a divided nation. And let that act of healing be the prelude to a lasting victory.' It was chilling, it was authoritative, and, after she had led the party for eight months, it was completely believable.

Soft and Warm to the Break of Dawn

If, in the mid-seventies, you happened to be listening to a late-night Washington radio station, and if you were in the required receptive state, you might have heard something you'd never heard before. Not on the radio, anyway. What you would have heard was the sound of sex, and in particular, Smokey Robinson's third solo album. With this record, he captivated listeners while inadvertently inspiring an entirely new genre, Quiet Storm. It would reinvent late-night radio, reignite the love ballad, and create a world where the soft soul version of love became an orthodoxy.

A Quiet Storm by Smokey Robinson

THE SEVENTIES PROVED TO BE an inevitable comedown for so many whose stars who had burned bright in the sixties. The decade had flattered novelty and celebrated attendance; by the seventies, as artists became established, so they became commonplace. All this new decade could offer was a career, rather than surprise. The seventies also unwittingly encouraged a host of imitators, opening up the market to a wealth of copycats who started to anticipate the moves of those they were employed to ape. Smokey Robinson was one such casualty, a man who, by the end of 1973, felt as though he didn't really know himself anymore, and who saw dozens of Smokey-lite wannabes wherever he looked.

He had been a Motown demigod, a man with a beautifully delicate high tenor who was a gifted songwriter: 'You've Really Got a Hold on Me' had one of music's greatest first lines ('I don't like you, but I love you'), while the lyrics to 'The Tears of a Clown' showed why

Bob Dylan had called him 'America's greatest living poet'. He had been on the road with the Miracles since he was sixteen, and he was tired. His wife, Claudette, had suffered seven miscarriages as a result of her being on the road, so when they did eventually have children (one through a surrogate mother), Robinson wanted to be at home. He said he didn't want to be absent, didn't want his children to be asking if they could have his autograph rather than saying 'Hello'. But he was getting testy. He was vice president of the label, he'd split from the Miracles, retired from performing and was terribly unhappy. He had tentatively embarked on what turned out to be a rather lacklustre solo career and was flailing. His first two solo albums (1973's *Smokey* and 1974's *Pure Smokey*) really weren't the hits he had been hoping for.

Elsewhere, as Eric Harvey pointed out in his seminal *Pitchfork* piece on Robinson, both Stevie Wonder and Marvin Gaye had become Motown's radical dynamic duo, attempting to turn the label into a platform for albums rather than singles. Both lyrically and musically, these men were exploring new territory and taking their audiences with them, reinventing the label for a different, more expansive, harder, socially aware decade. At the beginning of the seventies Wonder renegotiated his contract with Motown. He was in a strong position due to his almost unprecedented success and the new contract gave him carte blanche to do whatever he liked. No black American musician had ever been afforded such freedom and Wonder did not abuse it. Over the next few years, he recorded some of the most remarkable, as well as some of the most successful, records of the decade. Inspired by the likes of Marvin Gaye and Sly Stone, Wonder began mixing melodically sweet tunes with politically hard-hitting lyrics, exploiting the two predominant themes of black American music: the confrontational and the soulful. His socio-political lyrics were always accompanied by a sweet refrain ('You'll excuse me,' he once said, 'I don't consider myself to be a black performer. I consider myself to be a performer who is black'). Making full use of electronic wizardry (clavinets, synthesisers etc.) he created his own version of progressive rock, toured with the Rolling Stones and soon became the pre-eminent black entertainer of the day.

Motown even came up with a slogan for this reinvention: 'If you liked the boy,' they declared rather clumsily, 'you'll love the man.' His appeal was universal: he was favoured by the mass middle-of-the-road audience as well as the white critical establishment who acclaimed him as an innovator and social commentator. The albums came fast and furious. First there was *Music of My Mind* (1972), then *Talking Book* (1972), *Innervisions* (1973) and *Fulfillingness' First Finale* (1974). After this, he renegotiated his Motown contract again (it reputedly ran to 120 pages), with a 20 per cent royalty rate and a thirteen-million-dollar advance. He was bigger than Isaac Hayes, George Clinton and Sly Stone, bigger even than James Brown.

Marvin Gaye had become a black superstar too and with his 1971 concept album *What's Going On* created one of the pivotal song cycles of the decade, a fusillade of social realism that is spiritual in tone yet journalistic in execution. The narrative established by the songs is told from the point of view of a Vietnam vet returning to his home country and witnessing hatred, suffering and injustice. Gaye's introspective lyrics explore themes of drug abuse, poverty, ecology and war.

The seventies marked a significant shift in the soul music landscape, as artists began to experiment with new sounds and styles. At the time, the title track was the most avant garde single the label had ever released, while the album itself saw Gaye stretch himself as a producer, songwriter and singer. 'I felt like I'd finally learned to sing,' he told biographer David Ritz. 'I'd been studying the microphone for a dozen years, and I suddenly saw what I'd been doing wrong. I'd been singing too loud.' There is a languor to the record, a resignation that was evident in Gaye's performance.

Both men had reinvented themselves to such an extent that they were almost new artists, album artists who appealed as much to white adults as black ones. They were talked about as harbingers, visionaries and men with deep soul. But for the first four years of the seventies, Robinson had just looked like a sixties' throwback who wasn't going to last the distance. Robinson found Gaye's album especially vexing as he felt Marvin was deliberately trying to steal his thunder – not by his lyrical content, which was strident and political, but rather the production and what he called vibe. Gaye could sing about urban deprivation or the end of the world and still make it sound sexy. Robinson wasn't motivated to politicise his career, but he couldn't cope with being upstaged in the bedroom. He was also aware of the way in which Al Green, Barry White and Teddy Pendergrass were all having success with their own interpretation of sensual fireside soul music, none of whom were making records for Motown.

One of the reasons he ventured back into the recording studio was his signature, which he used hundreds of times a week in his position as vice president of Motown. He was signing so much he started using an abbreviation and then stopped altogether. He was desk-bound, unhappy and missed the creative spark of the studio. After about three years of corporate work, Berry Gordy, Motown's founder, came into Robinson's office and gave him his marching orders.

'He said, "Hey, man, I want you to do something for me." I said, "What?" So, I'm thinking he's going to tell me he wants me to go to New York or Chicago to make a deal with somebody, because that's what I did. This guy sits me down, and he says, "I want you to get yourself a band, I want you to make a record, and I want you to get the fuck out of here." He said, "Every day you come into this office, you are miserable." He said, "When I see you miserable, it makes me miserable and I don't want to be miserable. So I need you to get the fuck out." I could not believe this guy could see through me like that. I hugged him. So I went and wrote *A Quiet Storm*.'

A Quiet Storm was released on 26 March 1975, and it was something very particular. It had a mood all of its own, and an ambition to properly define the 'relationship' album. The album starts almost incidentally, slowly bringing you into a new world. Immediately the listener is transported to a world of questing and satisfaction, the two states of love. 'I finally had the musical concept I'd been seeking since hearing *What's Going On*,' Robinson said. 'I saw seven songs carried on the back of a breeze, blowing through the record from start to finish.'

The record opens with an arresting synthesised storm effect and an irresistible bassline, followed by close-miked, breathy vocals. The beat is barely apparent – we are a long way from Motown's trademark crashing snare here – but still, you find yourself getting lost in the groove, in its intimate life. The record was immediately soulful, absorbing, almost life-affirming. And steamy. It was certainly sexy, as was Robinson himself. The album became a template, a sultry soundtrack that played independently in the background of more pressing activities.

The title song was a statement. '"Quiet Storm" was my move back into show business,' Robinson told *Rolling Stone*. 'I figured I was a quiet singer and I said to myself, I'm gonna change my imagery and my vocal sound and I'm gonna take it by storm – quiet storm!' And so, he set to work, creating his own brand of contemporary, contemplative romantic soul. He would be an auteur of the night, an auteur of love. The resulting work was soulful, ambient, almost metaphysical. And it was a hit, appealing to an audience who were buying into Roberta Flack, Billy Paul, Jimmy Helms, Diana Ross, Minnie Riperton and, yes, Marvin Gaye.

In some respects, this was a strategically brave choice as the market was so crowded. There is no record of anyone using a collective noun for seventies soul crooners at the time, but an appropriate choice could have been 'boudoir'. They were everywhere: in addition to the likes of Marvin Gaye, Barry White and Teddy Pendergrass there was Bill Withers, Ron Isley, Curtis Mayfield, Bobby Womack, Lou Rawls and, of course, Isaac Hayes, the Casanova of cool. His 1969 album *Hot Buttered Soul* was a landmark, consisting of four psychedelic-orchestral tracks which were experimental, conceptual and introspective in design. Hayes whispered and rapped through his material rather than shouted, and he took ownership of other people's songs by being so confident and so particular. He sounded as though he were singing in a bedroom rather than a studio, and you could almost imagine him pulling back a red velvet curtain to reveal a room lit by candles and resonant with lust. His opulent versions of Burt Bacharach and Hal David's 'Walk On By' and Jimmy Webb's 'By the Time I Get to Phoenix' became instant classics, creating a new hybrid genre in the process – easy listening soul – he managed to subvert them by respecting their provenance. As for fashion, Hayes's signature outfit at the time was a gold-chained 'suit' which he said was a form of air-conditioning that helped him stay cool in the spotlight, as well as being a symbol of the end of black bondage and also 'a sex thing'. Every soul crooner had a sex thing, whether it was a come-hither medallion, a pair of spray-on slacks or just a guttural moan. Hayes took ownership of all around him, which was as much of a black power play as it was a land grab.

'Hayes took up time and space as if it was owed him, and listeners responded,' said the *New Yorker*. The album sold a million copies to black consumers alone. The record became so popular it changed the way FM radio saw the all-important weekend.

Barry White was perhaps the most unlikely soul star of the seventies: his lushly orchestrated soundtracks for seduction were delivered in a deep voice by a mountain of a man. Mocked by critics for his persona, which was a mixture of Isaac Hayes and Liberace, he created a fantasy world of opulence and desire. He was a love whisperer as much as he was a singer, crooning and croaking his way through 'Can't Get Enough of Your Love, Babe', 'You're the First, the Last, My Everything' and 'It's Ecstasy When You Lay Down Next to Me'. His concerts were as extravagant as his records, and at one point he was leading a sixty-two-piece all-girl orchestra. His formula was described as the simple translation of the paternal, consoling side of black preachers into the realms of love and sex, and it worked. On stage, in his bass-baritone growlings he would ramble on about the unity of love and the importance of relationships. The *New York Times* remarked upon his avuncular corpulence and his relaxed aura of control: 'Mr White is a smoothie – bland if you don't care, insinuating if you do.' His most popular nickname was the 'Walrus of Love' but he

also answered to 'Dr Love', 'Mr Love', the 'Prince of Pillo
'Ambassador of Romance', the 'King of Disco', the 'M
the 'Guru of Love'.

As for Sly Stone, even though he was still fully operational, he felt
as though he belonged to the past rather than the future. He may have
been held in high esteem by those people who knew he had appeared
at Woodstock and who understood his cultural importance and
resonance in the history of revolutionary political protest, but he didn't
seem very seventies. Sly and the Family Stone's fifth album, though,
There's a Riot Goin' On, released at the very end of 1971, was thought
by many to be their high point. It contained their very best single,
'Family Affair', a lo-fi examination of a couple trapped in a marriage
that their infidelities have rendered meaningless, which was 50 per
cent observation, 50 per cent lullaby. It became the biggest hit of their
career and surely few US chart-toppers could match its desperation
or bleakness. Stone would soon implode, unable to commercialise
his appeal or tailor his sound to the new decade. 'Family Affair' was
certainly sexier than he was, ironically one of the sexiest records of
the decade. It was a record that would become a staple in a new type
of black radio formatting.

Quiet Storm, the radio format, was born in Washington, at WHUR
96.3 – Howard University's contemporary radio station – when intern
Melvin Lindsey filled in as a substitute DJ one Sunday evening in May
1976, fourteen months after the release of Robinson's album. The
station had started to get a reputation for being a unique voice for
the booming black middle class in the DC area. Coming off the heels
of the civil rights movement, Washington, known as 'Chocolate City,'
had experienced a taste of black leadership with Walter Washington
serving as the city's first mayor. Dyana Williams, a radio host and
co-founder of the International Association of African American
Music Foundation, described the city as a black utopia. 'I had a black
dentist, a black gynaecologist and I banked at a black bank; everything
black, black, black, black.' And now they had a very particular kind
of black radio station, too. WHUR's FM signal reached further into

DC suburbs than the urban-centered AM stations did, and the less urgent programming style reflected this. One DJ even claimed on-air personalities were encouraged to 'sound less black' to appeal to middle-class suburbanites. The station's demographic ambitions were modified when the station hired sales manager Cathy Hughes to run it. She was a fan of 'psychographics', which focused on intangibles like attitudes and influences rather than age, race and location, and she used these metrics to analyse the station's listeners. What she discovered was a gap for a station that appealed to upwardly mobile, single, college-educated black women.

Hughes's idea would come to fruition by accident. Needing a last-minute substitute DJ for that fateful evening in May 1976, she brought in Lindsey, a Howard journalism student, WHUR intern – and her babysitter, to boot. He picked up some Detroit Spinners albums, along with a few by the Isley Brothers and the Delfonics and hurried to the studio. Somehow, despite his complete lack of experience, it worked. His confident, smooth but youthful voice was a contrast to the more aggressive black DJs of the day and his smooth, soulful selections immediately resonated with the listeners. Lindsey left the station to finish his bachelor's degree, but when he returned, eighteen months later, Quiet Storm quickly became a WHUR weeknight staple, and Lindsey became a local celebrity.

The format was simple and effective, not too different in structure from what album-orientated rock stations had been playing for years. Lindsey would play elongated stretches of uninterrupted slow tempo soul and R&B – sometimes up to an hour – only occasionally intervening to remind his listeners he was still alive. Essentially, this was private music, the kind you listened to alone, or maybe with one other person. It was . . . intimate. Right there in your home. Crystal-clear FM ruminations coming at you all night long. Sultry saxophones simmering over velvety basslines. Candlelit vocals caressing shadowy silhouettes swaying in embrace. This was mood music for romance. Quiet Storm normally consisted of ballads, slow jams, and sometimes sensual property.

'Lindsey's on-the-air personality is that of the soft-spoken alter ego for some 220,500 listeners each night,' wrote the *Washington Post*, 'not all of whom want to distinguish between radio and real life.' It wasn't unusual for listeners to call up and complain if the tempo sped up too much.

The tunes Lindsey chose weren't so much slow as glacial, the kind that should have been racked in record stores under 'Foreplay'. This wasn't joyous vulgarity, this was soul, and these were slow jams, every night a different sensual concept album. Roberta Flack. Marvin Gaye. Al Green. Donny Hathaway. Stevie Wonder. Millie Jackson. The Isley Brothers. And, of course, Smokey Robinson himself.

'You didn't even have to have a radio in DC,' said Donnie Simpson, the popular Washingtonian breakfast DJ. 'All you had to do is open the window. You couldn't help but hear "the quiet storm". That's how popular it was. It was the air that DC breathed.'

Lindsey created a mood, an ambience, a way to relax, driven by soulful, melodic mood music. As the radio format grew in popularity, it became a catch-all term, an umbrella term for industry ballads as well as new music which was being recorded specifically for this new genre. Lindsey became a genuine star, whose personality and calming voice contributed to a persona that would make him famous, not just in Washington, but all over the states. 'The next couple of songs I'm going to play have a special meaning for me,' he might say. 'I want you to listen to the words, as they're dedicated from me to you. And then I'll be back to say something before I go. But listen to these songs as they're just for you.'

The show initially ran only on the weekends, but after eighteen months it expanded to a five-hour nightly show. It became No. 1 in its time slot within a year. Until Lindsey's show, there was no place where you could listen to slow jams all night long, and he turned his show into a kind of virtual date night. He became so adept at mixing, and so subtle, that you often didn't realise when the songs were changing.

Quiet Storm DJs started to become famous; in the TV comedy series *WKRP in Cincinnati*, the radio station's own Quiet Storm DJ is called

DJ Venus Flytrap. Suddenly, slow, sultry soul was popular. Quiet Storm emerged at a time when the US black middle-class population was growing and the divide between the black rich and poor was widening. It was an escape from politics and friction and reassured black communities with a feeling of stability and normalcy. The 1999 book *What the Music Said: Black Popular Music and Black Public Culture* detailed why black people in the district were attracted to the format. Professor of Black Popular Culture at Duke University, Mark Anthony Neal, wrote that songs programmed for Quiet Storm were mostly 'devoid of any significant political commentary and maintained a strict aesthetic and narrative distance from issues related to black urban life.' The music was upscale, cool and mellow – 'all elegant and easy-flowing, like a flute of Veuve Clicquot champagne,' as *Rolling Stone*'s Ben Fong-Torres put it. There was some sniping, of course there was, as some critics complained it was a waste not to politicise Quiet Storm's success and use it as a platform for significant political commentary, but there were plenty of people who had been disappointed by the vagaries of publishing who were willing to embrace it.

The Myth that Jack Built

26 February, Salem, Oregon

BY 1975, JACK NICHOLSON WAS the Establishment's very own anti-hero, a seventies icon whose eyebrows were able to transmit everything from irony and violence to woe and licentiousness, often in the space of a single frame. He was his own rock star – a demented sex symbol, a rebellious spirit in a new iteration of Hollywood which was rebuilding its reputation with popular cinema verité, warts and all. Not that Nicholson had so many of those. When Easy Rider *(1969) took the sixties' counterculture overground and shook the Hollywood hierarchy, Nicholson started to enjoy artistic autonomy while selecting his roles with consummate skill. He wasn't shy about his talent either. 'I was at the Cannes Film Festival when they showed* Easy Rider,' *he said, 'and I knew what was happening. I was a movie star.' He figured he had waited so long to be famous that he deserved it. In Bob Rafelson's* Five Easy Pieces *(1970), his redneck oil-rigger suddenly revealed himself to be a lapsed concert pianist – an improbable twist that Nicholson made completely convincing. And in all his early seventies' performances – notably* Carnal Knowledge *(1971),* The Last Detail *(1973) and* Chinatown *(1974) – he combined irascible macho with an unexpected capacity for pain, tenderness and doubt. Which made him precisely the right actor at the right time. Jack could expose the fragility of machismo without looking weak, satirise the male mystique and still smell like a hero.*

He also understood his sexuality and his ability to use it. While the other icons of seventies' cinema (Paul Newman, Robert Redford and Clint Eastwood) were coy about their appeal, Jack positively revelled in it in a way no male star had done before. It was around now that the great apocryphal Nicholson stories started to appear, like the one of him at a European film festival. At a party following a screening, Nicholson is approached by a local newspaper reporter who had interviewed him earlier that day. She is, by all accounts, supremely attractive.

'Hi, Jack,' she says. 'Want to dance?'

Nicholson says nothing. He looks her up and down. Slowly, appreciatively.

'Wrong verb, honey,' he finally says. 'Wrong verb.'

When another reporter asked him about the veracity of the story, a few years later, he thought for a moment. 'Wrong verb? Well, I don't know if the story's true or not. It certainly sounds like something I would say. When I feel people are trying to provoke me, I sort of live up to it.'

By 1975, he had got to a stage in his career where he could drive people to the cinema to see work they weren't especially interested in, just to see Jack in it. He sought work and roles that didn't necessarily fit any particular genre. To him, genres felt old. And having spent most of the sixties playing bum roles in B-movies, he wasn't going to suddenly revert to type when he became famous. What Jack was going to do was what Jack was going to do, and if that meant appearing in difficult films, then so be it. We were still going to see them anyway. By 1975 he was, like Robert De Niro, a leading actor who didn't appear to be swayed by public opinion or indeed by the trappings of an industry which had spent a decade spitting him out. He had a big smirk, too, the kind of smirk that in the eighties would be appropriated by other actors playing for laughs. There was always something wicked about Nicholson's smirk, though, which is why it worked so well. Because seventies' cinema could still be extraordinarily wicked.

There were the shades, too, the sporting of Ray-Bans in inappropriate places, particularly indoors, and always at awards shows. Not many people could get away with this type of behaviour with their dignity intact, but with Jack it only seemed to add to the allure, add to the myth. In fact, there was no myth, but the actor had proved that he could bust out of nowhere, in an industry that he thought had already bid him farewell, and shine – splendidly – on his own terms. Jack Nicholson was perverse, decadent, and sometimes even a bit depraved. In other words, everything we wanted him to be. He knew that if he didn't act the way he did, we might be disappointed. He would even go out of his way to say that he only behaved the way he did because the audience expected it. Which we did.

One of Nicholson's best performances in 1975 was at the BAFTA Awards in London in February. Presented by a tuxedoed David Niven, in an almost full and woefully lit Albert Hall, he invited Twiggy to read the nominations for best actor. When she announced the winner, the telecast cut to a location shoot in Salem, Oregon, where Nicholson was filming Miloš Forman's One Flew Over the Cuckoo's Nest. *Based on Ken Kesey's book of the same name, it tells the story of Randle McMurphy who, in order to avoid hard labour after being found guilty of statutory rape, pretends to be mentally insane and is subsequently transferred to a high-security psychiatric ward dominated by the sadistic presence of nurse Mildred Ratched (played by Louise Fletcher).*

Nicholson stood behind a glass screen in his prison garb, another scene-stealing eccentric with more than a touch of lunacy. He started talking but quickly realised the screen was soundproof, so smashed it with his fist, foreshadowing Randle's eventual escape from the psychiatric ward at the end of the film.

'It's smashing to have been chosen as best actor this year for my performance in Chinatown,*' he deadpanned, while his fellow inmates (including Danny DeVito) formed a demented circle around him. 'I wish I could be with you there, but as you can see, I've been institutionalised here in Oregon,' he said, before being coyly led away by Fletcher. I saw the film at the Kensington Odeon one Saturday night, in early 1976, which was a formative experience as I remember finally feeling like an adult. Like most of my friends, I'd been sneaking into the cinema to see X-rated films ever since* The Exorcist *in 1973, furtively watching* The Decameron, The Night Porter, Last Tango in Paris, Flesh Gordon *and* Freebie and the Bean. *However* Last Tango *felt like a genuinely adult movie. I was an adult now, and I was going out to meet the world as well as waiting for it to come to me.*

The Boomers up on
Amsterdam Avenue

Like his apartment on the Upper West Side, Paul Simon was an
exercise in boho gentrification. His fourth post-Garfunkel album felt
as though it had been furnished rather than produced, sounding as
though no one involved in it had ever experienced the inequities of
youth. It was *so* grown up, like being in a Woody Allen movie or a
Neil Simon play. Paul Simon himself was becoming more considered
and the more careful he became, the more successful he was. It was
all about the ZIP code.

Still Crazy After All These Years by Paul Simon

STILL CRAZY AFTER ALL THESE Years was very much a neighbourhood record, and you heard it everywhere you went in New York in
1975. It was a Manhattan soundtrack, and in early October it spread
across the island like a contagion, from Wall Street all the way up to
Morningside Heights. In the Village, in Chelsea, in the forgotten parts
of SoHo which were slowly starting to be colonised and, in particular,
on the Upper West Side, this was what you heard.

In the seventies, the Upper West Side was the liberal heartland of the
city's intelligentsia. It was soft-belly Democrat central, Manhattan's
literary hub, and home to 101 bookshops. If you published a novel and
the launch party wasn't on the Upper West Side, the book world would
start to question your ambitions. The area had a soundtrack too
– Coleman Hawkins-style jazz, cabaret-style blues and gentrified folk-
pop, especially the kind produced by the likes of Simon & Garfunkel.

1975

If Woody Allen had started to co-opt the Upper East Side, giving every street east of Central Park a celluloid cameo, so Amsterdam Avenue and Upper Broadway continued to bathe in the mellifluous, supine sounds of 'The Boxer', 'Cecilia' and 'Bridge Over Troubled Water'.

By 1975, the Upper West Side had become the Greenwich Village of uptown, a place of egalitarian spirit, if not egalitarian wealth. It hadn't always been this way as, decades before, it was as rough as Hell's Kitchen, when school kids would leave the house with mugging money, just in case they got accosted. The streets then were full of genuine characters, as well as the sidewalk booksellers; one local volunteer teacher had a German shepherd that pulled a toy tricycle with a cat on the handlebars. But he was one of the more benign locals. The West Side of Manhattan was considered such a dangerously blighted area that invitations to parties on Riverside Drive were often rejected, large rent-controlled apartments were voluntarily given up, drugs deals were regularly conducted on brownstone stoops, vandalism was incessant and even pizza parlours wouldn't deliver above Columbus Circle. The neighbourhood started to change when the gay community moved in, particularly working-class men who were looking for their first white-collar jobs. The area's ethnically mixed gay population, mostly Hispanic and white, with a mixture of income levels and occupations, made it markedly different from most gay enclaves elsewhere in the city. The area also developed a keen sense of civic pride. This manifested itself at the tail end of the sixties, when Alexander's Inc., then one of the city's biggest retail stores, was set to open on Broadway between West 96th and 97th Streets. A two-year study helped Alexander's make its decision, its research showing that high-income individuals and families were moving in. But local resistance was huge, and they eventually withdrew their plans.

'The West Side is a community in which aloofness is considered a sign of weakness,' wrote Nick Pileggi for *New York* magazine. 'A gift horse on the West Side is considered Trojan until proved otherwise, and the announcement that Alexander's planned to wheel in a department store was taken by many individuals and neighbourhood groups as a declaration of war.'

Small businesses came back. The arts stayed along with its artists. The melting pot for which New York was renowned remained (and grew) as did the respective institutions, culture and cuisine. By 1975 it was a rent-controlled haven as well as being a real creative hotbed, with neighbourhood block parties full of literary heavyweights, poets, record-company bigwigs, journalists, children's entertainers, painters, academics and the flotsam and jetsam of the enfranchised classes. Manhattan at the time was experiencing a crime wave as economic stagnation was amplified by a large movement of middle-class residents to the suburbs, which drained the city of tax revenue. But the Upper West Side almost seemed immune to the chaos. And music was everywhere, pouring out of shops, cars, bars, restaurants and portable cassette players. And in 1975 you couldn't escape the sound of Paul Simon's new album.

Simon himself lived on Central Park West, the Upper West Side's golden ribbon. His beautiful duplex was furnished with Bauhaus chairs, expensive rugs, imported lights, an art deco piano and a Helen Frankenthaler painting. Neighbouring apartments had avocado-coloured floors, Laurel Canyon-inspired earth tones in the walls, colourful, geometric carpets and kitschy framed prints. Low-slung soft furniture accentuated the vaulted ceilings and brown appeared to be the colour du jour. The places to copy were Yves Saint Laurent's Paris library or Calvin Klein's Fire Island Pines home. Textured fabrics were everywhere, along with textured, geometric shapes and patterns and multi-use/free-flowing spaces like sunken living rooms, room dividers and upholstered seating. You couldn't move for velvet and rattan and patterned wallpaper. The Upper West Side was having a moment. When Simon left his apartment, he would be chauffeured by limousine for dinner at Elaine's, comparing private doctors and therapists with fellow diner Woody Allen (who would give Simon a scene-stealing two-minute cameo as a superficial, leisure-suit-wearing LA record producer named Tony Lacey in his 1977 movie,

Annie Hall). He was a very different man from the one he describes hitchhiking on the New Jersey Turnpike and boarding a Greyhound bus in Pittsburgh in his song 'America'. The characters in his new songs were also now a small world away from the students, bohos and beatniks who peopled the likes of *Bookends* and *Bridge Over Troubled Water*. Like many other musicians of his era, Simon suddenly had money. Real money.

Still Crazy was a New York record through and through, one that oozed wealth, style and uptown sophistication. The album was produced at A & R Recording, in midtown, the studio facility owned by the album's producer, Phil Ramone. While the production was less fussy than on *There Goes Rhymin' Simon* (his previous album), it had a studio gloss that unsurprisingly makes it sound more like a piano-centric Billy Joel album. That was entirely down to Ramone, who gave it a slick, Manhattan shine. It was a break-up record (like *Blood on the Tracks,* which was also partially steered by Ramone, albeit uncredited), and yet the production was so creamy that it never felt like anything other than a slightly maudlin skip down Fifth Avenue. It was lush when it could have been dour, sweet when it could have been sour. If Neil Simon had been a record producer rather than a playwright, he would have made records like this – metropolitan, knowing, a little sad, but with the most glorious peachy topcoat. The overall sound here was big and rich. There was texture to the instruments but a smooth quality to the vocals.

Even on his earliest records, Simon had been mature beyond his years, always sounding like a much older man. Here, he sounded almost sagacious. As he approached his mid-thirties – he turned thirty-four two weeks before the album came out – Simon was already adept at dealing with adult themes. And it obviously helped that he had just gone through a divorce. The tenor of his songwriting could be traced back to the gloomy opening line of 'The Sound of Silence' ('Hello darkness, my old friend'), but he had also managed to conjure just as many moments of levity and gratification. Exorcising personal demons and reflecting on midlife circumstance, Paul Simon framed

Still Crazy as a post-divorce statement, yet he retained his cutting humour and light-hearted cynicism that made every song universally appealing. Despite being capable of the extravagant and the wordy, Simon tended toward understatement, and his lack of vocal histrionics made his music seem deceptively effortless. It made *Still Crazy* so easy to hum as you walked along Broadway.

The photograph of Simon on the album cover was taken in the SoHo neighbourhood of New York with Simon standing on a fire escape at the southern end of Crosby Street, just north of the T-junction where it meets Howard Street. It was taken earlier in the year by Edie Baskin, who was dating Simon at the time, after his divorce from Peggy Harper, who he had married at the height of his fame in 1969. Baskin (the daughter of Baskin of Baskin-Robbins ice cream fame) was the official photographer for *Saturday Night Live.* Each week she took the shots of the hosts and music performers that would be hand-tinted and shown during the show's introduction and before and after commercials.

Paul Simon filled the album with such ennui it felt as though he was looking back at his life through a telescope. But then so many popular – and now moneyed – entertainers who had broken through in the sixties were already starting to look at that decade with rose-tinted glasses, as a sepia-coloured relic from their past which would forever hold them in its grasp. Musically, Simon had already started to deliberately mature by exploring indigenous genres, thinking that as his music became older it would be leavened by adopting gospel, reggae or Latin rhythms. Lyrically, he was as worldly as ever, as he had been writing about middle-aged angst from early times. This was the novelist's trick, acting considerably older than he was, and inflicting unearned wisdom upon his songs.

In the title song, when he talked about bumping into his old lover on a New York side street, it was novelistic rather than cinematic, a vignette that spoke volumes about his mood. This was a gentle, reflective song dwelling on middle age. Many baby boomers could relate to it immediately; Simon was born in 1941 and was growing older with his audience. He first played the song in full on the second episode of new comedy show *Saturday Night Live* (broadcast on 18 October), opening the show singing alone, sitting on a stool. As the song continued, the show's breakout star and original 'Weekend Update' sketch anchor Chevy Chase appeared as a stagehand, carrying equipment behind the singer. He fell over, looked up and announced, 'It's Saturday Night!'

Simon was the host that night, but despite the fact he wasn't even that episode's musical guest (who was Randy Newman), he still managed to perfume eight songs (including 'Me and Julio Down by the Schoolyard'). Notably, this episode marked the first televised reunion of Simon & Garfunkel since their breakup in 1970. Together they performed 'The Boxer' and 'Scarborough Fair'. Much has been made of the animosity between them, but here they seemed happy, with Garfunkel putting his hand on Simon's shoulder and Simon gently reaching back to him. When Simon returned to the show in November the following year, he once again opened the show with 'Still Crazy . . .', this time wearing a Thanksgiving-appropriate turkey costume.

Simon had debuted the song much earlier in 1975 on *The Dick Cavett Show*, an appearance that became instantly famous because the song was only half-finished. Though Cavett didn't seem to understand the uniqueness of the situation.

When he had started writing the song, he knew it was special, but immediately felt compromised. 'Sometimes the process can be uncomfortable, as the songwriter is forced to confront aspects of his own life he'd rather avoid altogether,' he said. 'But in the service of the song, such sacrifices get made. The title came out of nowhere. I knew it was song-worthy, but I wasn't crazy about what "Still Crazy" told me about myself.' Some years later, he would trace the album's title song to a particular moment when he felt overwhelmed by sadness, summarised by Robert Hilburn in his 2018 biography of Simon: 'How, after all his success, could he feel so bad? Suddenly, a phrase came to him that would summarise his feelings. Ironically, the phrase, born in sorrow, would be the title of a song [and album] embraced by millions as an upbeat expression of survival.'

This luxurious introspection was very a la mode. Me-ism was everywhere in this part of New York. Bioenergetics, TM, e.s.t., rolfing, Gestalt therapy, vegetarianism, tai chi, zen, yoga, acupuncture, psychoanalysis, the lot. Self-scrutiny, self-awareness and self-fulfilment. All the selfs. 'Me' was big business in America, and what swung on the West Coast, principally in the greater San Francisco area and extremely specific area codes in Los Angeles, swung up here on the Upper West Side. Of course, the big idea was meant to be a move away from communitarianism (sixties), towards atomised individualism. People were wary of branching out alone to too great a degree, and so consequently little subsects sprung up, groups of people who felt safe in each other's company, and who consumed the same food, movies, magazines and media. And music: *Still Crazy* was a serious common denominator for those at the sharp end of Me. The seventies were all about a new alchemical dream: changing one's personality – remaking, remodelling, elevating and polishing one's very *self*. More and more boomers were leaving college and settling down with families of their own. They did not have time for marches against the war in Vietnam and, besides, it was already winding down. Americans were turning inward, seeking comfort in spiritual renewal or seeking insight by visiting therapists, reading self-help books, or exercising. They also sought insight through music, particularly music that mirrored their own experiences. The seventies were in many ways a decade of fads and crazes. Whether in fashion, exercise, literature, play or dance (disco, say), people picked up new activities and products with abandon and dropped them soon after. But they clung to music that 'spoke' not just to them but for them.

When Simon was interviewed by the *New Yorker*, back in 1967, when he was twenty-five, he spoke about his audience, who, he figured, were already the age he would be in 1975. 'I can look into people and see scars in them,' he said. 'These are the people I grew up with. For the most part, older people. These people are sensitive, and there's a desperate quality to them – everything is beating them down, and they become more aware of it as they become older. I get a sense they're thirty-three, with an awareness that "Here I am thirty-three!" and they probably spend a lot of afternoons wondering how they got there so fast. They're educated, but they're losing, very gradually. Not realising, except for just an occasional glimpse. They're successful, but not happy, and I feel that pain. I'm drawn to these people and driven to write about them. In this country, it's painful for people to grow old.'

He went on to describe how a thirty-three-year-old woman might feel, knowing she was no longer seventeen, or how thirtysomethings were rearranging their priorities by stopping smoking, getting a suntan, going on a diet, playing tennis etc. He could have been talking about the people buying his records in 1975. But of course, he was much better at writing songs for them by then. 'What's intriguing is that they are just not *quite* in control of their destiny,' he'd told the *New Yorker*. 'Nobody is paying any attention to these people, because they're not crying very loud.

Still Crazy After All These Years is not a long album: its ten songs are pretty much all in the three-to-four-minute range, for a total running time of just thirty-five minutes. Yet it was incredibly varied, literate and insightful, with a broad synthesis and songs in a variety of styles and genres, with a unifying sense of orderly assembly. It all works. Side one kicks off with the title track, which is suitably downbeat, but so elegantly painted. There follows the Simon & Garfunkel reunion on 'My Little Town'; the doleful little 'I Do it For Your Love'; '50 Ways to Leave Your Lover', which was a huge international hit (Simon said the rhyming gambit was copied from a game he was currently playing with his three-year-old son, Harper); and 'Night Game', ostensibly an account of a baseball game. Side two then socks you with the gospel of 'Gone at Last', featuring Phoebe Snow (after Simon and Ramone had initially envisioned a Latin-tinged duet with Bette Midler), before almost rushing through such delicate songs as 'Some Folks' Lives Roll Easy', 'Have a Good Time', 'You're Kind' and 'Silent Eyes'. It all felt so self-contained and perfect, as though it wasn't possible to improve it. Yes, you could have played it differently, and certainly produced it differently, but it seemed as though it was just about as good as it was possible to be. With his tender, conversational voice, his almost mathematical arrangements and suitably mellifluous production, Simon had created an emotionally complex masterpiece.

Paul Simon was an adult when he made *Still Crazy*, and yet he remained obsessed with his youth. 'I'm always trying to make "Mystery Train" [by Elvis Presley] – it's my favourite record,' he said. 'And that's what I mean. You get this sound in your head at some early point in your life, and you love it so much. Then you spend your whole career trying to make sounds that work.'

Tommy, Can You See Me?

18 March, Broadway, New York

WILLIAM FAULKNER PUT IT BEST in his novel, Requiem for a Nun, *when he wrote, 'The past is never dead. It's not even past.' For Pete Townshend, the creative force behind the Who, after* Tommy, *the past would never leave him. Before, his band used to end their concerts with 'My Generation', which had always been epochal in its anthemic value – once they started playing it, the audience could do nothing but explode. But after* Tommy *– his genre-defying rock opera about a deaf, dumb and blind kid who discovers a messianic gift for pinball – they started performing 'See Me, Feel Me' and 'Listening to You' at the end of their set, which they'd play until the audience got on their feet. It wouldn't matter how long it took because Townshend knew that eventually they would all stand up. 'Because it's like a prayer,' he told me, 'an incessant prayer, a secular prayer.'*

Rock's first commercially successful rock opera became the band's calling card, a piece of work that was ripe for constant reinterpretation. There was always going to be a film, and in 1974 Townshend allowed Ken Russell to start making one. This was a big deal, as it signalled another significant phase in the development of post-war pop culture. You could tell rock was growing up as people were starting to make 'popular' movies about it. Not countercultural films, not glorified biopics or cinematic documentaries, not the kind of stuff aimed at fans, but films about pop that were going to be watched by the kind of people who went to the cinema, the kind who were going to see Chinatown, Bring Me the Head of Alfredo Garcia *and* Young Frankenstein.

Ten years previously, A Hard Day's Night *had been, like the endless list of Elvis movies, a way to engage and exploit the Beatles' fan base. It didn't really matter that it turned out to be a masterpiece, as the audience would have gone to see it anyway.* Tommy *was something else entirely;* Tommy *was a grown-up film based on a genuine piece of adult-orientated rock. Not for the kids, ma'am, not for the kids!*

Only it just happened to be directed by a lunatic in Ken Russell, whose over-the-top provocations were excessive by any measure. He had already taken great liberties with his portrayals of classical composers, making his films highly sexual, baroque extravaganzas. But while he had built his reputation on developing movies which were deliberately a bit nuts, nothing ever topped his psychedelic interpretation of Tommy.

Russell had already made Women in Love *(which featured an infamous nude wrestling scene between Alan Bates and Oliver Reed),* The Music Lovers *(Russell said if he hadn't told United Artists it was a story about a homosexual who fell in love with a nymphomaniac it might have never been financed) and* The Devils *(featuring Vanessa Redgrave passionately kissing Jesus) but he appeared to have found a perfectly designed project in* Tommy. *Townshend's masterwork might have been full of excesses, but they were nothing compared to Russell's. The movie, which had the structure of a vaudeville show, was laced together with some magical vignettes, not least Ann-Margret rolling around in a sea of baked beans wearing a spandex catsuit before throwing a champagne bottle into a TV set (literally being assaulted by the products of consumer culture). The first time the actress met Russell she thought he was like a lamb, all meek and diffident with bright-blue eyes and a mischievous glint. He was rather different on set, as his obsessive tendencies took over. The baked bean scene took three days to film, with take after take after take. Each time he wanted another shot, Ann-Margret would have to go back to her trailer, change, wash her hair, dry it, curl it, and come back as though nothing had happened. The crew members were all wearing high boots to protect them from the mess, although Ann-Margret's catsuit shrank a little each time, making the shoot even more sexually charged. Plus, the music was so loud that she couldn't hear any of Russell's directions. By the end of the shoot, she was covered in cold baked beans, blood and broken glass. To me she was a cinematic wet dream, a slightly tarnished Atomic-era sex kitten who was obviously prepared to subvert her image for the sake of Hollywood glory. Even though I was a massive Who fan, I was far more interested in gawping at Ann-Margret swimming in a sea of beans than looking at Roger Daltrey in a bath full of the things (which is what I'd been doing ever since buying* The Who Sell Out*). Russell had originally told her that her character was having a nervous breakdown, and that she could do whatever she wanted – which is actually what she did. She was nominated for an Oscar, though, and the film was a massive hit.*

By 1975, Townshend had got used to the new parameters of his industry. In 1969, when the Who had sold a million dollars' worth of Tommy *albums, the* New Yorker *accompanied them to a ceremony and press conference in the grand ballroom of the Holiday Inn on New York's West 57th Street, where they were going to be given a bunch of gold records (they were in town for a week, performing the album at the Fillmore East). The assembled hacks were keen to quiz the band on their experiences at Woodstock, as the US tabloid press at the time still had a tendency to treat British rock stars as imported curios and Woodstock, which had disconcerted the country because of the apparently effortless way it had galvanised a generation, was of particular interest. Townshend (described by the* New Yorker *as thin, with large hands and wrists, a small chin, and an oversized nose), perhaps predictably, was dismissive in a way that only a British rock star could be.*

'I don't think we'll be playing gigs like Woodstock in the future,' he said. 'A festival is essentially a public occasion and the music is just background.' He went on to mention a new festival which he said was going to be even worse. 'One promotor I've heard about is trying to organise a festival in Wyoming next summer. Not any particular part of Wyoming – the whole state.'

1975

With Tommy, *the movie, Townshend was reaching the masses without compromising, at least not the kind of compromising he was talking about. By 1975, he was already in discussion about making a movie of the Who's 1973 album,* Quadrophenia.

A Broken Piano and the
Future of Jazz

By the early seventies, jazz had veered off into the world of fusion, leaving the path free for more accommodating styles. To wit: for the residents of Cologne, the evening of 24 January wasn't destined to be a memorable one and yet it proved to be magical for the 1,400 people who saw Keith Jarrett perform a solo recital in the gilded surroundings of the city's famous opera house. It was the first ever jazz concert performed there, and its recording – released in November – became the new jazz benchmark.

The Köln Concert by Keith Jarrett

THE HISTORY OF SUCCESSFUL LIVE recordings is a litany of serendipity. It's the accidental cassette recording of a small club gig; a bootleg that was never officially meant to see the light of day; the unforeseen final gig; the unexpected first gig; the only gig. Classic live performances are rarely orchestrated; when the Who recorded *Live at Leeds* in the refectory of the city's university on Valentine's Day 1970, they didn't know it would be hailed as one of the greatest live recordings of a seventies' rock band; ditto the MC5 and 1969's *Kick Out the Jams*. The Velvet Underground classic *Live at Max's Kansas City* was recorded at the famous nightclub and restaurant at 213 Park Avenue South in New York in August 1970 on a cassette tape casually left on a table by Brigid Polk, one of Andy Warhol's acolytes. It was only when the record company scenester Danny Fields asked Polk to rewind the tape that anyone knew it had been recorded for posterity. Five days later Lou Reed left the band.

The same fortuity occasionally happens in jazz. Friday, 24 January 1975 proved to be something of a memorable night for the audience that saw the American jazz pianist Keith Jarrett perform a solo recital in the courtly surroundings of the imposing, modernist and incredibly prestigious Cologne Opera. It was an extraordinary building, an imposing, austere auditorium designed and opened in 1957 by the famous German architect Wilhelm Riphahn, who also built the city's circular Restaurant Bastei. Its interior was similar to the Festival Hall in London, built six years earlier. Tickets were four Deutsche Marks, which was a lot at the time, but by the time he arrived for his concert, Jarrett had already built a considerable reputation.

The native of Allentown, Pennsylvania was in the middle of a European tour, travelling between gigs in a battered old Renault 4, with one of his record label's founders and his manager, ECM's Manfred Eicher. They would release the evening's recording as a double album wrapped in an uber-cool and rather stately monochrome sleeve, with a deliberately impassive photograph of Jarrett. The whole mood was dispassionate and detached, much like the label itself, which very firmly placed itself in the vanguard of contemporary jazz. In some ways the label was quite austere and, while the record certainly didn't look particularly commercial, it would turn out to be a massive hit; it had a meditative, spiritual and transformative power, and quickly established itself as a classic, selling four million copies (it's still the all-time best-selling piano album). Remarkably, though, the concert very nearly didn't happen.

A precocious seventeen-year-old German student, part-time promoter and jazz obsessive called Vera Brandes was responsible for organising the concert and, at Jarrett's request, had arranged for a ten-foot, eight octave Bösendorfer 290 Imperial concert grand piano to be provided. Jarrett had felt he had been nagged into appearing and, although he didn't want to let this novice of a promoter down, by asking for this particular instrument he was giving himself some security: the Bösendorfer originated in Vienna early in the nineteenth century. It is said to be the first concert piano able to stand up to the playing technique of the young virtuoso, Franz Liszt, whose unforgiving treatment of pianos tended to destroy them. But when Brandes took Jarrett out onto the stage for the rehearsal to meet the piano, it turned out to be a rehearsal model, with nothing grand about it at all. It had keys that kept sticking, some totally broken, and the felt on others was worn away so the upper register sounded very harsh and tinny. The pedals didn't work. And because it wasn't a grand piano, it wasn't loud enough. Jarrett, a perfectionist, was furious. It was a 'piano which hadn't been adjusted for a very long time and it sounded like a very poor imitation of a harpsichord or a piano with tacks in it,' he said. He started playing it, stood up, circled the instrument, sat down to play some more, and then silently stood up again. After standing in silence for nearly a minute, he spoke to Eicher, who then spoke to Brandes.

'If you don't get another piano, Keith can't play tonight.'

And that was that.

'I fell from the sky,' Brandes told the BBC years later. 'I just didn't know what to do. So I ran down to the phone booth outside and started calling anyone I knew who might be able to get a substitute piano. I also called a piano tuner to come and see if he could fix the bad piano. Honestly, if the concert hadn't gone ahead there would have been a riot. It would have been impossible to give everyone their money back.'

Jarrett was not in the best of moods. He had played the Salle des spectacles in Épalinges, Switzerland the day before and had made the sleepless, four-hundred-mile journey overnight. They had been sent airline tickets but decided to cash them in and drive instead – fully aware of how demanding his concerts could be, Jarrett had insisted that they schedule a concert every other day. He was also experiencing appalling back pain and was wearing a brace. His state of mind wasn't helped by the fact that the concert wasn't due to start until 11 p.m., the only time Brandes could secure, following that evening's opera.

After calling around, she soon found out that another venue in the city where she was hoping to promote a concert had the right piano, and so she started rounding up enough friends to help move it. Then it started to rain. By then the piano tuner had arrived and demanded a great deal of money to move the backup piano carefully to avoid it being ruined by the rain. 'The piano tuner saved our lives,' says Brandes, who then begged Jarrett to commit to the gig as he sat in his car ready to head back to his hotel. Standing in the rain she pleaded through the open window of his tiny Renault. Jarrett looked out at the bedraggled teenager and took pity on her. After a few moments of silence, he said, 'Never forget. Only for you.'

He was twenty-nine, and yet suddenly felt a lot older. Exhausted, back at the hotel he tried to nap but couldn't manage it. He and Eicher went out to eat, according to Jarrett, 'in the hottest Italian restaurant I've ever been in, and I was sweating profusely. We were sitting with about ten people, and everyone was served before I was. My food arrived fifteen minutes before I was supposed to be at the hall, and I had to gulp down food that was not very good in an overheated restaurant, having not slept for twenty-four hours.'

Tired, still hungry and lumbered with a suboptimal instrument, the piano player abandoned any preconceived ideas of performance and adapted to a new reality. This in itself was strange, as he didn't like audience distractions, especially when he was improvising; he had started to hand out cough drops to audience members during winter concerts, and he would sometimes play in the dark to prevent people from taking pictures. Tonight, he was more equitable. 'I was forced to play in what was – at the time – a new way,' he said. 'Somehow, I felt I had to bring out whatever qualities this instrument had . . . my sense was, 'I have to do this. I'm doing it. I don't care what the fuck the piano sounds like.'

He did what he had to do, not because he thought it would be good, but because he felt he had no choice. In effect he did what his old boss Miles Davis had told him to do: 'Don't play what's there. Play what's not there.' 'Because he could not fall in love with the sound of [the piano], he found another way to get the most out of it,' said Manfred Eicher.

While his style had to change for the performance, the content did not; like every concert on this tour, every show was improvised, a massive undertaking for any jazz musician, especially one performing solo. As he sat down at the piano he was so tired that he momentarily closed his eyes, but then jolted himself into a performance that shocked even him. The opera house had a signal bell, a cadence of five chimes that told patrons to take their seat. When Jarrett started he played the five-note sequence, and as he did, a ripple of laughter ran through the auditorium. Then he developed that same musical idea a bit further . . . and he built a twenty-six-minute improvisation around the motif. The audience immediately knew this was going to be a bespoke performance. 'Just as quickly,' wrote Corinna da Fonseca-Wollheim in a 2008 *Wall Street Journal* article titled 'A Jazz Night to Remember', 'the reaction turned into awed silence as Mr Jarrett turned the banal and the familiar into something gorgeous and mysterious . . . In the jazz world of 1975, the sheer beauty of the programme was revolutionary.'

What he did was very particular. Unlike the European 'kings of improv', Jarrett was obsessive about playing traditionally, while his own improvisation was based on just a couple of chords. For instance, in 'Part I' he spent almost twelve minutes playing over Am7 and G. Despite these limitations, the audience were rapt, transfixed, and the atmosphere in the auditorium soon became church-like.

They were watching a genius at work, a man possessed. Jarrett overcame the piano's lack of volume by standing up and playing it extremely hard. He bashed the keys as he stood, sat, moaned and writhed, grunting to himself like a tennis player. The groans and sighs from Jarrett are obviously spontaneous, and it's almost as though the music is coming from way down in his gut and he is forcing it out of his body towards the keys. You can feel the intensity, as he is totally transfixed in himself and his connection to his instrument, even though it's not an instrument he has any affinity with. It's a musical version of his battered Renault. On the record you can hear the agony of the music, and his effort at creating any sound at all. He sounds like more than one man up there on stage; in fact, he sounds like a whole band. Afterwards, he would say he had spent the first twenty minutes working out which were the good keys, and which were the duds, focusing on only certain parts of the keyboard in an attempt to add some coherence to his work.

In 2010, an anonymous blogger, who had been working for Charisma Records at the time, posted this online: 'In 1975, I happened to be in Cologne for reasons I can't recall, and heard about the Keith Jarrett concert from Vera Brandes, the very youthful promoter, who pulled off a major coup in getting Jarrett in the first place and then overcoming several serious setbacks to stage the concert. It didn't start until nearly midnight and, as I had only found out about it at the last minute, I didn't get a seat, but stood just inside one of the many entry doors of the opera house. It didn't matter though, because I was so entranced by Jarrett's performance that I didn't even notice. My main memory is of his intensity . . . he moaned, groaned, swayed, stood up, sat down, stood up again . . . segueing effortlessly from passage to passage, seemingly oblivious to his surroundings and the enraptured audience. Time just flew by and suddenly it was over . . . the audience, who were certainly aware that they had witnessed something really special, were applauding wildly, relishing their good fortune at having witnessed it . . . and I have to say, I was too.'

When he arrived in Cologne, even though Jarrett was a young performer, he was seasoned, with ten years of experience. He'd already played piano for saxophonist Charles Lloyd before joining Miles Davis's band in 1970, when the trumpeter's music was becoming unprecedentedly dense, expansive, and rhythmic, with songs often stretching to the half-hour mark, featuring multiple bassists and drummers. It felt chaotic, progressive, and appealed to jazz longhairs. The two Miles records that included Jarrett – *Live at the Fillmore* and *Live-Evil* – are frenzied and aggressive, much like Davis was at the time, as one critic put it, 'Jazz as Afro-futurist squall, a Jackson Pollock deconstruction of Duke Ellington's famed "jungle sound" from the twenties.' It soon became too much for Jarrett, who needed a different type of crazy and a different kind of solace. 'Freebop' was never going to do it for him. Jazz in general felt exhausted, almost like classical music, having reached both an endpoint and a cul-de-sac. There didn't seem like there was any road left to explore; it was an art form that was no longer popular. Pop was popular, rock was popular, but not jazz. You could freeform all you liked, but even that felt like an act of appropriation, a jazz version of jamming, like Jimi Hendrix, the Allman Brothers or the Grateful Dead. After all, if jazz wasn't modern, then what was it? Jarrett left Davis at the end of 1971, renouncing both electronic instruments and jazz-rock.

He was soon approached by Manfred Eicher, who had recently started ECM. Eicher promised artistic freedom, management, and money, which resulted in his first solo album, *Facing You*, made up of solo improvisations. This was a different kind of jazz, being 'clean, direct and acoustic'. Jarrett was influenced hugely by the great Bill Evans, probably the single most important figure in 'jazz without swing', whose impact can be traced to one recording: the brilliant six-and-a-half-minute solo piano improvisation called 'Peace Piece', recorded in 1958 for his album *Everybody Digs Bill Evans*, the year before he joined Miles Davis to record *Kind of Blue* (closing track 'Flamenco Sketches' was heavily influenced by Evans's composition). 'Peace Piece' used the opening chords of Leonard Bernstein's 'Some Other Time' as a repetitive figure over which Evans improvised, creating a classic in the process. If there is any single piece of music that could be cited as a source for Jarrett's solo concerts it would be this.

Facing You proved to be popular, which meant that Eicher encouraged Jarrett to go out on the road to promote it, in Europe as well as the US. He would become the new face of acoustic-jazz-as-concert-art music. And he soon became a crossover star, popular for his athletic stagecraft as much as his music, standing up at the piano, wildly contorting his arms, breathing heavily, and weirdly chanting along with his melodies. Which is why he could sell out a 1,400-capacity opera house at eleven o'clock at night.

After the concert, Jarrett didn't seem to understand what he had wrought during that frantic hour on the stage, and it wasn't until later in the car on their way to the following day's concert that he and Eicher listened to the cassette of the concert. What they heard confirmed their belief that the evening had been special, a serendipitous occasion that was worth memorialising. He later explained what had happened: 'It just seemed like everybody in the audience was there for a tremendous experience, and that made my job easier.' In spite of some reservations about the technical quality, the recording was to be released.

The Köln Concert has spawned all kinds of columns, talks, books and BBC and NPR documentaries, over the years, and it's not hard to understand the enduring appeal of the story. The economist Tim Harford gave a TED Talk about adversity as a catalyst for creativity and focused on the concert as a case study. The concert was a personal triumph, a professional tour-de-force and a musical benchmark. And it came out of nowhere.

'In many ways, he's a very freethinking musician,' said Harford later in an NPR documentary with Guy Raz. 'He does these entirely improvised concerts. He sits at the piano, and just plays whatever comes into his head. So that sounds as though this is a very loose, flexible person, but he has a reputation as insisting everything be perfect. So, the perfectionist meets the world's worst piano, and it's a sell-out. Everyone's going to be there in a couple of hours. There is no way to get a replacement. And Jarrett said, I won't play. So this is not going well for Vera. So she goes out. She finds Jarrett. He's sitting in a car waiting to be driven back to the hotel. And she knocks on the window, and he looks out, and sees this seventeen-year-old kid drenched in the rain, and she just begs him to play. I think at that moment, he just feels sorry for her. He realises she's just a kid. Fourteen hundred people are about to show up and there's going to be no concert. And he says, "Never forget. Only for you." And he agrees to play. [And] within moments, it's apparent he's producing something astonishing. It was supposed to be a disaster. He's given this unplayable piano. And he doesn't just cope. He doesn't just produce a decent performance, because he's a genius. He produces what many think of as his best performance. I think [the audience] were spell-bound. Whether they knew it was particularly unusual – maybe they thought it's always like this – I don't know. But certainly, the music has stood the test of time.'

Harford goes on to say that Jarrett's instinct was probably wrong, and that in hindsight he should have perhaps rolled with the punches and gained a bit more appreciation for the unexpected advantages of having to cope with a little mess, but that doesn't seem to be the moral of the story at all. Jarrett's reluctance to embrace the situation started in motion a sequence of thoughts which caused him to develop a way to cope with it. This wasn't petulance, it was alchemy.

'The minute he played that first note everyone knew this was going to be magic,' said Brandes. 'It's something I will never forget. The first note and everybody was totally mesmerised. He was like an acrobat. He moved very intensely. And he was singing in a very special way. You can hear that in the recording, and you can hear he's stomping with his foot. This was a very physical experience. That was very special, and it was very special that night. He was really in it.'

The Köln Concert quickly became one of those jazz records enjoyed by people who don't like jazz, like Miles Davis's *Kind of Blue,* Dave Brubeck's 'Take Five', pretty much anything by Astrud Gilberto or the kind of sixties' loungecore albums that are regularly used now for electric car commercials. And while it's totally improvised, it bears little of the unwieldy hallmarks of Ornette Coleman or Thelonious Monk. It was also quasi-classical and decidedly European, making European jazz a genuine force in the global jazz market. For others, Jarrett's work became a gateway drug, opening up the seventies' sophisticate to an exponential world of dissonance and wonder. Getting into jazz can be like suddenly discovering you have an extended family you knew nothing about, while the family in question runs to thousands of members. Like turning the world upside down and finding another one underneath, a world where they only ever listen to jazz. *The Köln Concert* was also a fine example of where tastes were in 1975, a manifestation of deluxe sophistication.

Gentrified. Clipped, contained, framed. It was certainly a world away from Bourbon Street.

There were many who thought Jarrett was black or biracial; he had dark skin, wore a moustache and an Afro, although in the mid-seventies this wasn't unusual. The American was in fact of Scots-Irish and Hungarian descent, and was middle class. Some were critical of his success, saying that he had reinscribed the problem of authenticity coupled with the notion of privilege. But then Jarrett wasn't alone in this.

'Jarrett, as a kind of miscegenated presence, in effect legitimized white jazz as something that does not swing but that is just as much jazz as its black counterpart,' wrote Gerald Lyn Early in 'Keith Jarrett, Miscegenation and the Rise of the European Sensibility in Jazz in the 1970s', in 2019. 'The fact that there was such ambiguity about Jarrett's race and that he performed this type of music through a European record company may have had much to do with his success. There was something about this music coming from Europe that gave it a certain gravitas and something about this music coming from someone whom many people thought was black.'

In the early sixties, Thelonious Monk used to play regularly at the Blue Note on 3rd Street, New York. He used to rehearse in the afternoons, a half-eaten sandwich left on top of the piano. Bob Dylan once dropped in to see him play and told him he played folk 'up the street'. 'We all play folk music,' replied Monk.

Pinteresque

IT WAS BLEAK, PARADOXICAL AND enigmatic. *When Harold Pinter's* No Man's Land *opened, at the National Theatre at the Old Vic, starring John Gielgud and Ralph Richardson (and directed by Peter Hall) in April, there were many who thought it represented a weird kind of projection of his own darkest fears, a nightmare vision of the isolation of fame. After all, he had been notable (if not famous), for nearly twenty years and it didn't always sit well with him.*

One summer evening, two sexagenarian writers, Hirst (a wealthy recluse), and Spooner (a down-at-heel poet), meet in a Hampstead pub and continue their drinking into the night at Hirst's stately house nearby. As the pair become increasingly drunk, the conversation turns into a war of attrition and an increasingly revealing power game – do they know each other or is each performing an elaborate character of recognition? – further complicated by the return home of two sinister and acerbic younger men, the one ostensibly a factotum, the other a male secretary (sidekicks, essentially). Slowly, their relationships are exposed, with threat and hilarity (Pinter's plays have been famously described as 'comedies of menace'). Like all Pinter's work, it was almost unbearably claustrophobic (someone had the audacity to call it a Beckettian parable about old age), with all four inhabiting their no man's land between time present and time remembered, between reality and imagination – a territory that Pinter explored with his characteristic mixture of biting wit, aggression and anarchic sexuality. Dark? Yes, it was dark. Spectral? Yes, that too.

The Guardian's *Michael Billington, who at the time was just about the most important theatre critic in the country, said the play 'is about precisely what its title suggests: the sense of being caught in some mysterious limbo between life and death, between a world of brute reality and one of fluid uncertainty. But although plenty of plays, from* Sweeney Agonistes *to* Outward Bound, *have tried to pin down that strange sense of reaching into a void, I can think of few that have done so as concretely, funnily and concisely as Pinter's.'*

His world was usually defined within the confines of a single room, seedy in his early work, and later slightly more elegant. The play was unique in as much as the conflict appeared internal, and laden with memory.

There were those who thought Pinter was a travesty of a person – one critic called him 'insufferably pompous, sanctimonious, philistine, humourless, politically sciolistic, intellectually suburban' – and yet it would be difficult not to call him the greatest British dramatist of his generation. He wrote the kind of plays that actors loved because he wrote monologues for them and allowed them to achieve greatness just by being stared at. Even when they weren't speaking, they were thinking. But Pinter was incredibly thin-skinned and didn't respond well to criticism, something exploited by the satirist Craig Brown, who teased him mercilessly, not least with his spoof poem, 'Polly Put the Kettle On' ('Polly put the fucking kettle on!'). Once, Brown saw him across a crowded room at a party. After the playwright made a throttling gesture at the satirist, the hostess said to him, 'Do you want to punch him?' and Pinter said, 'I wouldn't dirty my fists.'

It was a busy year in the West End, with the opening of The Black Mikado *(adapted by Janos Bajtala, George Larnyoh and Eddie Quansah from the Gilbert and Sullivan),* By Jeeves *(written by Andrew Lloyd-Webber and Alan Ayckbourn) and Stephen Sondheim's* A Little Night Music. *London's theatreland had been transformed since the removal of the Lord Chamberlain's office of censorship in 1968; the Arts Council, which was quite a radical body at the time, started investing in small-scale theatres and alternative theatre had started to blossom in universities. Theatre groups were springing up all over London, driven by a sense of radical cultural insurrection. This was the underground at work, a community forged in the pages of* Time Out, *in the bars of Soho, and the townhouses of Islington. This was the golden age of the fringe, which saw the growth of the socialist theatre movement in Britain and the rise of companies such as The People Show, Pip Simmons, CAST and Red Ladder. There was a groundswell of activity, with dozens of performances each night, small troupes deliberately pushing the boundaries of what used to be considered acceptable. If the sixties continued to manifest itself in any tangible way, exploring the parameters of sex, race, and politics, it was generally to be found in the fringe theatres of London. Here, on the margins, every venue was a Petri dish, every play a manifesto, every performer an ideologue; there were possibilities literally everywhere, and if it wasn't always love at first sight, it was certainly fascination.*

London was full of disrupters, although there was no greater disrupter than Pinter. The actor Douglas Hodge appeared in fifteen of his plays and knew Pinter well. He told the Guardian: *'All you ever heard about Harold was him trying to attack people or fight people, but he was this fantastically complex person. He was the most violent pacifist I ever met, but that's why he was an artist, not a politician. Whether he was right or wrong about his beliefs, he was always certain. He was also loyal like nobody else. I remember, when we were doing* Moonlight, *talking to the stage-door guy, who could only have been about seventeen. He had the job of going round this massive, spooky theatre at night to lock up, and I offered to give him a hand one time. He said, "Well, I have to get it done quickly because I'm going down to the pub with Harold Pinter." I said, "What do you mean?" And he said, "Well, I told him I'd written a play, and he read it, and he made some notes."'*

What Price an Ambient Notion?

Could music ever sound like an art school? Could it be vertical as well as horizontal? Why weren't recording studios considered to be musical instruments? Sonic oddball Brian Eno had started to make a career of asking himself – and others – seemingly difficult questions. And now he was making a song and dance about turning his work into conceptual art. There was no expectation he was going to be a star when he left Roxy Music, but by 1975 he had established himself as a sculptor of sound.

Another Green World by Brian Eno

ON THE EVENING OF 18 January 1975, while walking back in the drizzling rain from a session with his old Roxy Music colleague Phil Manzanera at Island Records' Basing Street studio near Notting Hill's Portobello Road, on his way home to Maida Vale, Brian Eno had something of a premonition.

This was pre-gentrification Notting Hill, when the area was still a marginal wasteland, before the developers and estate agents moved in and before its carnival became Europe's biggest street party (in 1974 it had entertained just over 100,000 people). It was also slightly rough, so walking home to Maida Vale – thirty minutes away – was perhaps not Eno's smartest ever idea. But Eno was apprehensive. Not that this was particularly unusual. For about a week he'd been sensing that he was about to have some kind of accident. It was the same feeling he'd had before he'd got appendicitis when he was sixteen, and the same feeling he'd had before his lung collapsed when he was doing his solo tour, the previous year. He always seemed to sense when he had pushed things too far, too hard. It happened whenever he had been carried off by the momentum of media approval or professional opportunism and had ceased to think about where he was or what he was doing. He had just finished a song called 'Miss Shapiro' (which Manzanera would release later in the year) and had found himself asking what he would feel like if this happened to be the last thing he ever recorded. Was it good enough? He thought to himself, Well, probably not. He felt dark, and rather morbid. Then, immediately, he started to think, What the hell are you talking about? You know, what a ridiculous train of thought to be on.

Two minutes later he slipped on a rainy pavement and stumbled in front of a taxi. The taxi was doing about forty miles an hour and, although Eno stepped back instinctively at the last minute, it was too late. It smashed into him, running over both legs and throwing his body back against a parked car, so he hit his head. Eno felt blood running down his face and, when he touched the top of his skull, it felt as though it was split wide open. He wrapped both hands around his head, not knowing what condition it was in. Some bystanders quickly called an ambulance, although they didn't think he would make it.

Meanwhile, Eno knew he had to get up, as he could see another car coming around the corner, following the taxi, at approximately the same speed. He remembered worrying about who was going to get in touch with his girlfriend, and then thinking about what a wonderful organ the brain is, able to have all these different thoughts at once. The next thing Eno knew, he was being wheeled along a hospital corridor, both hands still wrapped desperately around his skull.

'The whole thing was horrible,' he said. 'I was conscious all the way through. And I was thinking, You stupid cunt, you brought this on yourself.'

Weeks later, as he was lying in bed at his flat in Maida Vale, he had a lightbulb moment. His friend Judy Nylon had been to visit him, bringing an album of eighteenth-century harp music. Eno felt too weak to get out of bed and turn the volume up. As he lay there, his mind wandering, his body aching, he heard a strange convergence of sounds, of barely audible music and a light rainfall outside. He said he could hardly hear the music above the rain – just the loudest notes, like little crystals, sonic icebergs rising out of the storm. At first, he wished he could turn the damned thing up, but then started to think how beautiful it all sounded. As he listened, he decided that this 'melted-into-the-environment' quality was what he wanted in his music and in his life. What if, he thought to himself, you could make music that was designed to be heard but not actively listened to? 'This presented what was for me a new way of hearing music,' said Eno, 'as part of the ambience of the environment, just as the colour of the light and the sound of the rain were parts of that ambience.'

He set to work making *Discreet Music*, an ambient album that would eventually come out at the very end of December. He had invented what would be known as the genre of ambient music. 'Ambient' did, of course, have a classical music forerunner in the *musique d'ameublement* ('furniture music') of Erik Satie – tinkling sketches meant to decorate a room – and of course in the work of John Cage ('4'33", for instance, which is four and half minutes of silence), but it was Eno who introduced that minimalism into the context of contemporary rock music. Eno was already a big fan of Cage, having been introduced to his work by his art school professor, the painter Tom Phillips (whose painting Eno would crop for the cover of *Another Green World*); it was then that he started to understand that making music didn't have to be dependent on playing an instrument. And he was going to put it into practice. These were proto-chill beats, sounds you could pay attention to or choose not to be distracted by it if you wanted to do something while it was on. Eno was anticipating that music was becoming so embedded in most people's lives that was rarely the centre of attention. It really was the happiest of accidents.

This work also encouraged Eno to launch Obscure Records, an experimental label which would become a vehicle for even more esoteric recordings. He was robust in his reasoning, using Mike Oldfield's *Tubular Bells* as an example. 'What's happening – though the record companies haven't yet realised this – is that people are listening to music in a lot of different ways and for a lot of different reasons,' he told *Impetus* magazine. 'There are a number of different instances of this which, regardless of what you think of them musically – whether you like or dislike them – stand as interesting examples of this. For example, *Tubular Bells* is a highly successful record, and that indicates that people are quite capable of sitting and listening to an uninterrupted twenty-five-minute piece of music which doesn't have lyrics and doesn't have all the things that were thought to be necessary to sell a record. And there are other instances like Tangerine Dream; the Terry Riley records have sold very well, terribly successful records those. Velvet Underground's "Sister Ray" – an incredibly good seller again. Some of the German bands, some of the Floyd's things.' He was starting to realise that there was a sonic world that was relatively unexplored, a world that grew out of the possibilities of recording and performing. It started to become clear to him that there was a new form of music evolving – a music not necessarily based on performance: music where the compositional emphasis was as much on texture or timbre as it was on the more familiar areas of melody, harmony and rhythm.

Having recorded *Discreet Music* and having successfully launched Obscure Records, it was perhaps not surprising that by the summer, when he was recording what was meant to be his third 'proper' solo album, he decided he was sick of the sound of his own voice. Ensconced in Island Studios, he was finding it difficult to wrap his voice around the material he was producing. Six months later, in November, when the record was finally released, only five of its fourteen tracks would feature vocals, a sharp contrast with his previous albums. He could have separated the vocal tracks from the instrumental ones, as David Bowie would do a couple of years later on *Low*, on which Eno worked, but instead he dispersed them at even intervals, 'like lily pads of song between deeper seas'. He said the effect was like slipping into and out of sleep while friends talk in the next room.

He had viewed the record as something of an experiment and had entered the studio with nothing prepared. For the first four days, he produced precisely nothing, and so turned to his instructional cards, the Oblique Strategies. He had recently created these with multimedia artist Peter Schmidt, and they took the form of a deck of cards offering gnomic suggestions intended to help artists (particularly musicians) break creative blocks by encouraging lateral thinking. Whenever a musical or production challenge arose, a card would be drawn, providing cryptic guidance or direction. The instructions included things like 'Abandon normal instruments', 'Ask people to work against their better judgment', 'Change instrument roles', 'Discover the recipes you are using and abandon them', 'Don't be frightened of clichés', 'Do something boring', 'Faced with a choice, do both', 'Make an exhaustive list of everything you might do and do the last' and 'Use "unqualified" people'.

Eno had used Oblique Strategies on his previous album *Taking Tiger Mountain (By Strategy)*, but this time he was going to push the envelope. Consequently, some of the album credits for the instruments have fanciful names that describe the sound they make – 'Castanet Guitars' were actually electronically treated mallets; the 'Leslie piano' was an acoustic piano miked and fed through a Leslie speaker with a built-in revolving horn speaker; elsewhere there was a 'snake guitar' and an 'uncertain' piano.

The Genesis drummer Phil Collins was asked to contribute, although not in the traditional fashion. '[Eno] gave us all a bit of paper, and we made lists from one to fifteen,' said Collins. 'Eno said, "No. 2, we all play a G; No. 7 we all play a C sharp," and so on. It was like painting by numbers.' The drummer got about twenty beats in before stopping to throw beer cans at a bicycle across the room. Even Eno himself got so frustrated that he would come home from the studio and cry, later calling the process 'almost unmitigated hell'.

King Crimson guitarist Robert Fripp, who had worked with Eno on their collaborative album *No Pussyfooting*, in 1973, was also seconded. Eno asked him to improvise a lightning-fast guitar solo that would imitate an electrical charge between two poles on a high-voltage generator. Which to him didn't seem like an unreasonable request.

He was fond of saying that it would make a huge difference to the approach of recording if there was a sign above every studio door saying, 'This Studio is a Musical Instrument'.

Eno gave a very good impression of appearing to be more interested in machines than people. He kept referencing the back cover of dub album *King Tubby Meets the Upsetter*, on which there was a photograph of the mixing consoles instead of the artists. Eno wasn't entirely prescriptive though and would hire musicians (or ask them to work with him) based on what they could deliver creatively, rather than simply do his bidding. The idea came from a remark Robert Wyatt made to him about how Miles Davis did his arrangements: he didn't rely on an arranger but instead he'd ring up a guy in New York and a guy in Paris and a guy in Los Angeles, tell them to be at such and such studio at a certain time and take it from there.

'Art is automatically subversive in some sense,' Eno told *Time Out* when the album was released. 'I don't mean that art is going to lead to Marxism, or anything as distinct as that. But the function of art is to keep you ready to innovate.'

Eno became obsessed with challenging what happened in recording studios, bored by the accepted orthodoxies. He liked to describe the sensation of driving a car towards a set of traffic lights. If the light turned green, he would say, you would think nothing of it and drive on – there would be almost no information because it would be expected. But if the light turned blue, a considerable dilemma would arise. A mental procedure would take place which would force you to revise your whole view about traffic lights. If the traffic light moved towards you and tapped the side of the car, you might reject the information completely. Eno operated on the notion that there was a certain threshold at which you would be absolutely incapable of classifying what's going on and so you would make no attempt to. It was important that the information appeared to use a language you understood, but in a way, perhaps, that you also didn't understand. But if the language changed entirely, like the traffic light started walking, then the tendency would be to reject it entirely. So, his view of music was that the interesting things were the ones that were on or very close to the threshold.

To create the lyrics on the few songs he did sing on, Eno would play the backing tracks and then sing nonsense syllables to himself over the top, before turning them into actual words and phrases and meaning (he was convinced quite a lot of Bob Dylan's *Blonde on Blonde* had been written that way). What Eno was inventing was a new kind of pastoral, ambient pop. On his previous records, it was almost as though he was attempting to reimagine glam rock as sound sculpture, which fitted in with his previous role in Roxy Music; on this one, he was deliberately moving into new territory. With Roxy, he had been allowed to develop the persona of the mad professor, but when his ideas started to irritate Bryan Ferry – who was regarded as the chief architect of Roxy – Eno found himself surplus to requirements. Never mind, he thought, I'll go off and make weird records by himself. Which he did, and which were surprisingly popular. *Another Green World* was another story.

At the time he said he was no longer concerned with making horizontal music, by which he meant music that started at point A, developed through point B and ended at point C in a kind of logical or semi-logical progression. What was more interesting was constructing music that was a solid block of interactions. This then left your brain free to make some of those interactions more important than others and to find which ones it wanted to speak to. In Eno's world, vertical music was 1975's big idea. One thing he liked about it was the fact that you could enter it at any point and leave it at any point. You didn't have to start at the beginning. Like watching a film on constant repeat, dipping in at any moment. He was also experimenting with making a combination of horizontal and vertical where it would be possible for both of them to exist at once.

As *Pitchfork* so cleverly identified a few years ago, one of the most remarkable things about the album was how a stoic Englishman who showed no interest in the conventional expression of emotion managed to make something that felt so intensely personal. 'I read a science-fiction story a long time ago where these people are exploring space and they finally find this habitable planet,' he said to the *NME*, reflecting on the album's title. 'And it turns out to be identical to Earth in every detail. And I thought that was the supreme irony: that they'd originally left to find something better and arrived in the end – which was actually the same place. Which is how I feel about myself. I'm always trying to project myself at a tangent and always seem eventually to arrive back at the same place.' As the title suggests, there was a recurring theme of nature and environment throughout the album. Tracks like 'In Dark Trees' and 'Another Green World' evoked images of calm, serene landscapes, untouched by human interference. The *NME* described the album as 'a journey into the very essence of sound'.

The album's title track had already been heard on television. In October, the BBC launched a flagship arts documentary series, *Arena*, founded by Humphrey Burton, who at the time was head of the BBC's music and arts department. Even at the time the strand felt uniquely transgressive, and over the next few years would empower the likes of Alan Yentob, Nigel Finch and Anthony Wall to produce some of the best arts documentaries ever made: *My Way*, *The Private Life of the Ford Cortina*, *Desert Island Discs* and more. The title music, accompanying film of a message in a bottle, which at first felt almost incidental, was 'Another Green World', a deceptively simple, circular theme of great beauty.

Eno was himself the subject of a 2010 *Arena* film subtitled, obviously, *Another Green World*, although he asked for another piece of music to be used over the credits. At one point he complained that the amount he was paid in royalties every time the theme was played on television was rather meagre: 60p. Long before the days of Shazam, *Arena* received countless letters from viewers about the music. The producers started collecting them and when they moved from their offices in 2013, they shared some of them in a six-minute film, which was itself a very *Arena* thing to do.

I'd loved Eno since *Taking Tiger Mountain (By Strategy)*, loved the way everything he did was so weird and oblique. I knew he'd been kicked out of Roxy, knew the backstory, and enjoyed the way in which he was determined to pursue a solo career. He was a constant feature in the *NME*, a genuine maverick who appeared to have a social life almost as complex as Bryan Ferry's. Unusually, he also seemed to have a sense of humour, which was strange for someone who liked to present himself as an artiste. One of his favourite jokes, which I heard him tell many years later, concerned a Texas rancher who decides to return to the UK to rediscover his English heritage. He visits his great uncle, a Somerset farmer, and asks him how far his land extends. 'All the way from the top of the hill right down there to the street,' replies the yard farmer, gesturing proudly. The Texan chuckles. 'I can get up in the morning, drive all day long and not reach the end of my property,' he says, smugly. His uncle replies, 'Yeah, I used to have a car like that.'

While *Another Green World* was venerated by critics, it failed to chart in either the UK or the US. It nevertheless charted a course for Eno's entire career, the not-so-mad scientist using the recording studio as a musical instrument; indeed, using anything as a musical instrument, sometimes even humans. Eno would go on to make music for films, music for airports, for art galleries, civic recovery centres and, inadvertently, music for spas.

The Fall of Saigon

30 April, Ho Chi Minh City, South Vietnam

WHEN NORTH VIETNAMESE TROOPS MARCHED into the capital on 30 April 1975, it marked the most crushing defeat in US military history. Even though the Vietnamese joked that the communists took Saigon 'without breaking a light bulb', casualties were heavy on both sides, although the fighting stopped just short of the city limits. Throughout all the years of conflict, war had not often touched Saigon, with the exception of the occasional random rocket attack, a restaurant bombing or two and the dramatic but limited incursion into the city.

The event marked the end of the Vietnam War and the collapse of the South Vietnamese state. In the days before, US forces evacuated thousands of Americans and South Vietnamese. American diplomats were on the frontline, organising what would be the most ambitious helicopter evacuation in history. They landed at ten-minute intervals at the US embassy, including on the roof. With some pilots flying for nineteen hours straight, more than seven thousand people were evacuated, including 5,500 Vietnamese.

The conflict was the definitive proxy combat of the Cold War. Ultimately, the US and its partners failed in their objective to stop the perceived threat of the 'Red Menace'. The war was also the most contentious political conflict of the age, and its influence on contemporary culture was huge, forming the basis of hundreds of books, films, plays and musical anthems. Unlike the two world wars, which inspired morale-boosting culture; the Vietnam campaign produced a countercultural insurgency that concentrated on its atrocities, the opposition to the war and its lasting effects on society. The noise was everywhere, from Country Joe and the Fish's 'I-Feel-Like-I'm-Fixin'-to-Die Rag' in 1965, to the tsunami of Hollywood films through the seventies and eighties. Some of the most noteworthy include Michael Cimino's The Deer Hunter *(1978), Francis Ford Coppola's* Apocalypse Now *(1979), Oliver Stone's* Platoon *(1986) and Stanley Kubrick's* Full Metal Jacket *(1987). Other Vietnam films include* Hamburger Hill *(1987),* Good Morning, Vietnam *(1987), Casualties of War (1989),* Born on the Fourth of July *(1989),* Forrest Gump *(1994) and* We Were Soldiers *(2002).*

As it came only three years after the fall of Saigon, Cimino's film was immediately divisive. The film's grim depiction of young Americans' loss of innocence stirred raw emotion in audiences in 1978, three years after the war's end. In one famous scene, captive soldiers are tortured and forced to engage in a grisly game of Russian roulette.

Ned Tanen, then the president of Universal Pictures, felt the emotion at a preview showing in Detroit. 'The screening was a blood bath,' Tanen said. 'Some people were very moved, but we had walkouts and lots of people who were enraged and totally shocked.'

He was confronted by a burly auto worker who stormed out of the theatre. 'He said, "Do you have anything to do with this [expletive] movie?" And when I said, "Yes," he grabbed me by the throat and wrestled me to the ground ... I mean, this is the same film that won the Oscar, but here's this guy trying to punch my lights out in the lobby.'

The response to The Deer Hunter *reflected the divisions the war wrought in American society. For young and old, rich and poor, and for all races, it became the great divide, the catalyst for a seismic shift in contemporary pop culture. 'Vietnam' was in a category of its own. This war, once one among many, became The War – the war that divided the world's greatest superpower, set the rest of the world afire, and shaped an entire generation.*

The denouement was precipitated by Watergate and Nixon's resignation in August 1974. After that, Congress reduced military aid, more than mirroring public opinion – during the conflict, more than 58,000

Americans had been killed and more than 300,000 wounded. South Vietnam had fallen to the communists and the war had made Americans question the integrity of their own governmental institutions.

The legacies of the war were already in play. If the Vietnam War killed trust, it helped with racial integration. By 1975, there had been over a quarter of a century of integrated service, and the hundreds of thousands of servicemen and women who came home from Vietnam had different ideas about race — some for the worse, but most for the better. There was a broad consensus among professional historians that the Vietnam War was effectively unwinnable, although it became something of a template for most interventionist US wars going forward. It wasn't just the American public and its culture that questioned America's involvement in Vietnam. The memory would start to strain the country's ties with some of its traditional partners in Western Europe, as there was now always the fear that America would allow itself to be drawn into 'another Vietnam'.

The war also managed to create a generation of heroin addicts. The proximity of Vietnam to the Golden Triangle, the area overlapping the mountains of Myanmar, Laos and Thailand, made heroin easily accessible during the war. It was also a convenient alternative to weed. After John Steinbeck detailed the GIs' rampant marijuana use in the Washingtonian *magazine, there was a huge public outcry. This prompted a policy shift by the military, resulting in stringent penalties for marijuana use. As a consequence, many soldiers shifted to heroin, which was easier to conceal. For returning GIs, addiction was often as traumatic as PTSD. Predictably, conspiracy theorists liked to say the military and intelligence agencies were complicit in the drug trade. Reports suggest that US planes, including the CIA-operated Air America, would return from delivering supplies to anti-communist forces in Laos carrying enough heroin to feed an army.*

Twenty-Three Orgasms in Seventeen Minutes

This is the record that in many ways portended the eighties, a club song that was remixed and extended to fuel a sexual fantasy but in reality, created a format that dance music has adhered to ever since. 'Love to Love You Baby' fills the whole first side of Donna Summer's second studio album. Clocking in at sixteen minutes and fifty seconds, it quickly became both a dance club favourite and a fashion boutique soundtrack as well as the template for the 12-inch disco remix.

Love to Love You Baby by Donna Summer

AT THE BEGINNING OF FEBRUARY, in the middle of a snowstorm, Giorgio Moroder was fast asleep in his Munich apartment when the phone rang. He'd been working late in his studio, Musicland, which he had built in the basement of a hotel outside the old Munich airport. He'd been tinkering with the Donna Summer record he'd been making and, having got home late, had fallen asleep almost immediately after hitting the pillow. Moroder was exhausted, it was the middle of the night, but he couldn't ignore the phone. He picked up the receiver to hear the pinched, whiny voice of Casablanca Records bigwig Neil Bogart on the other end.

'Giorgio? Hi. I want you to make it longer,' Bogart said, forgoing any pleasantries. He was talking about the song Moroder had played for him a few days earlier at the Midem new music convention in Cannes: a slow, sexy but rather sombre track Moroder had co-written with its singer, Donna Summer.

Born in Italy, Moroder had briefly studied architecture before deciding to pursue his passion, playing guitar, starting to tour Europe in various cover bands. He moved to Berlin, then Munich, where he got into production and where he opened his studio with Pete Bellotte, a Brit who would become his assistant. He started playing around with synthesisers and developed his own sound.

'In 1970 an engineer I knew, called Robbie, introduced me to a classical composer in Munich who had this incredible new instrument,' he said. 'It was a humungous machine with cords everywhere and he played me this composition which just consisted of a bass tone that kept changing every half minute. That was his composition! He was using this huge machine to create what was known as "musique concrete". There were no rhythms, no effects, and it wasn't too interesting. But then, when he wasn't around, Robbie took me aside and said, "Look, with this synthesiser you can create more than just a low note." He showed me a few things and I thought, Wow, this is great!' Immediately fascinated by the machine's possibilities, he started to play around in the studio, and soon recorded a synthesiser-based song called 'Son of My Father', which was covered by the British group Chicory Tip.

Moroder had also become intrigued with *Switched-On Bach,* which had been released in 1968 by the American composer Wendy Carlos. This was a collection of pieces by Johann Sebastian Bach, performed by Carlos and Benjamin Folkman on a Moog. By the time Moroder heard it, the album had nearly sold a million copies.

Moroder and Bellotte looked for a singer to record a demo for the group Three Dog Night. A friend recommended a twenty-year-old American, Donna Gaines Sommer, a Boston native who had moved to Munich to star in a local production of *Hair.* Moroder thought the new girl had a terrific voice and asked her to record a song called 'The Hostage', which became a European hit. When the record company misspelled her last name as 'Summer', Moroder insisted she keep it.

'I really just thought of "Love to Love You Baby" as a bit of fun,' said Moroder. 'I had some free time one afternoon in the studio, so I started playing around with Donna's idea. The way I demo'd back then wasn't much different to the way I work now. In 1974 the first cheap little drum machines came out, so I would use one of those, and I also had a real drum loop with several different tempos. I would put up a tape from a twenty-four-track and I would have a mic for the vocal, as well as some sort of keyboard – a Fender Rhodes, maybe. Having established the tempo of the song, I would just record the rhythm, along with a guide vocal, and then go from there.'

Fascinated by its sensual falsetto and languid pace, Bogart had gone crazy for 'Love to Love You Baby', brought it back to Los Angeles, and played it during a party at his Beverly Hills mansion. The party turned into an orgy, fuelled by vast amounts of cocaine, a drug that had renewal as its DNA, the repetitive lust for continual sensation. People kept bumping into the record player, jogging the record back to the beginning; Bogart noticed that every time this happened, people cheered. It was a genuine lightbulb moment.

'How long do you want it to be?' Moroder asked Bogart.

Bogart said he wanted the song to last the entire side of an album – sniff – and then said Moroder should go and listen to Iron Butterfly's mad, seventeen-minute 'In-a-Gadda-Da-Vida'. 'That's what I want it to sound like,' said Bogart. 'Well, actually I don't want it to sound anything like that, but I want it to last seventeen minutes.'

Which is what Moroder did.

In the morning, he told Summer about Bogart's request. 'Giorgio, how am I supposed to come up with lyrics for another fourteen minutes of the song?' she asked. The producer, who had been thinking about this all night, had a solution. Later that day, down in the studio, he lit a few candles, dimmed all the lights, and told Summer to improvise. 'And make it sound sexy,' he instructed her. Summer certainly did, and soon the sighs and moans of a woman having an orgasm became the narrative. This new version became a breathy homage to Serge Gainsbourg's infamous 'Je t'aime . . . moi non plus,' where the French pop star could be heard pretending to fuck the actress Jane Birkin. Moroder had asked Summer to think of a persona while she sang, and she chose Marilyn Monroe and her whispered rendition of 'Happy Birthday' to President Kennedy during his forty-fifth birthday party at Madison Square Garden in 1962.

'Just the concept of having something new like that – nobody had one whole side of an album with one song,' said Moroder. 'Secondly, it was relatively easy. And the fact that it was seventeen minutes was the reason for its success. It was played in discotheques and especially in the gay community they LOVED it. I loved the whole thing.'

This was now a record that mirrored excess: you could dance all night to it, have sex all night to it, float above the clouds all night to it. It was an automatic, perpetual soundtrack. And it would change dance music for ever.

'I was using my voice more as a sound instrument than as a singer, almost like using myself like a synthesiser,' Summer said. 'Giorgio, Pete, and I were enamoured with what we could do with sound. I don't think we knew that we were forging a new pathway. We were just being ourselves.'

After the sixteen-minute, forty-nine-second version was released, no one apart from Neil Bogart thought the song would get airplay. But New York DJ Frankie Crocker started spinning it after midnight and encouraged others to do the same. Nightclub DJs picked up on it and suddenly 'Love to Love You Baby' was a hit. Bogart called it 'a beautiful, great balling record', telling people to 'take Donna home and make love to her – the album, that is', and encouraging radio stations to play the track at midnight as a catalyst for home listener romance ('Tune in for twenty minutes of love'). The song caused a commotion, and famously inspired a ban by the BBC, whose censors somehow calculated that the track contained twenty-three orgasms. Regardless, it broke the Top 5 in both the US and the UK, aided by Casablanca Records' decision to aggressively market its singer as 'the first lady of love'. She was photographed wearing slinky satin clothes, holding her microphone as though it were a phallus. Little was left to the imagination. For some it seemed like a sexed-up novelty single, something exacerbated by the sultry schlock of Summer's live performances – she'd often be carried onstage by two men clad in loincloths, while her backing singers simulated sex behind her. The audience went delirious, taking their clothes off, throwing underwear at the stage and dancing partially naked.

The highly eroticised music was fundamentally at odds with Summer's background – as a child, her father smacked her for wearing red nail polish because, he said, 'That's what whores wear.' Her family were deeply religious, and this completely informed her life. Summer spent much of her early life in church, where she first began to sing. As she sometimes struggled to hit the high notes, one day she prayed, 'God, please teach me how to sing better.' Church was her source of faith and hope. Crucially, it was also the place where she was abused at the age of fifteen, something she kept secret for decades. Being reinvented as a sex symbol didn't sit right with her, and she soon started saying she was uncomfortable with her new image. 'I'm not just sex, sex, sex,' she told *Ebony* magazine. 'I would never want to be a one-dimensional person like that.'

Nevertheless, a sex symbol was what she was, labelled 'the Black Panter', 'the Linda Lovelace of pop', 'the sex goddess' and 'the queen of "Sex Rock"'. In 1975, Vince Aletti was a critic working for the *Village Voice* and the music trade magazine *Record World*. In the 20 September issue of the latter, he wrote 'The most talked about album cut right now is Donna Summer's extraordinary "Love to Love You Baby", the title track and entire side (16:50!) of her debut album . . . She does little else but moan passionately and repeat the title . . . it had everyone racing to the booth, asking, "What is that?"' He was talking about the New York disco The Loft, in Prince Street in SoHo, which started to play the song with such regularity it almost became an anthem.

Disc jockeys all over America were learning the same tricks, which put a premium on those doing it well. If you were inventive, the crowd came back. Aletti described David Mancuso's style, one of the DJs at The Loft, as a tapestry, a constantly surprising mix of sound that manipulated the crowd in a way he'd never experienced before. He said it became addictive.

'I remember them playing "Love to Love You" in the club, the first time for a club crowd, and just being amazed that it could work the way it did – that it could be that overwhelming and sexy and danceable at fifteen minutes plus,' he said. 'It had that orgasmic sound to it; I don't remember a song before that proceeded so much like lovemaking. Donna's voice was so thoroughly woven into the music, but it was really this overall sense of eroticism that made it work. It had the length that you could really flow with it and let it completely take you away. A lot of DJs had mixed songs that would approach fifteen minutes of length, but to start with something that was around fifteen minutes was pretty incredible at that point. The fact that it became a radio record was really a surprise to anybody involved; I think initially most of us thought it would just be a great club record. A kind of breakthrough to the power of disco at that point, that radio found the record so compelling that they were willing to play the whole thing. The length, the sensuality, the storytelling, and the sense of a new sound coming up – it really did start there with Donna Summer. It became a very specific sound that eventually took over – it became what people thought of as disco. And then, when disco was over, that was what people stopped hearing.'

Two years later, Moroder and Bellotte would reinvent dance music again when they produced Summer's 'I Feel Love', the first song to combine repetitive synthesiser loops with a continuous four-on-the-floor bass drum and an off-beat hi-hat, which would become the defining characteristic of both house and techno ten years later. The only non-electronic instrument on the record is the bass drum, played by future Billy Idol producer Keith Forsey. At the time, they were experimenting with genre songs for Summer, celebrating the forties, fifties, sixties and early seventies. When they decided to make a track that actually sounded like the future, they came up with 'I Feel Love'. It would turn out to be the most prophetic thing they'd ever done. With its locomotive bassline and pulsing synths, it sounded like it might never stop. This was metronomic relentlessness. When Brian Eno first heard it, he was living in Berlin with David Bowie, working on their own version of the future. Having found a copy of the twelve inch, he rushed into the Hansa recording studio and said, 'This is it, look no further. This single is going to change the sound of club music for the next fifteen years.'

Moroder's breakthrough didn't just influence a new generation of electronic sound, it influenced traditional dance music, too. 'Giorgio had funky, R&B kind of roots, mixed with a vision of the future,' said Nile Rodgers. 'It sounded clever and innovative, almost on the level of jazz pioneers like Charlie Parker. When I wrote "I Want Your Love" for Chic, that was me trying to imitate on guitar what Giorgio and these guys were doing with keyboards. I said, "This is the funkiest, tightest person that has ever walked this planet." I didn't think that any musician could play with that kind of virtuosity. I had no idea that there were computers running it. I was shocked.'

He wasn't the only one.

South of Houston

2 May, Manhattan

BY 1975, THERE WERE MANY in the art community who thought that New York was very much now in the shadow of Los Angeles. If downtown Manhattan had been at the forefront of artistic endeavour during the sixties, spearheaded by Andy Warhol's almost wholesale adoption of life below 14th Street, by the mid-seventies the city had experienced a creative exodus. Los Angeles had the heat, it had the sun, and it had a sense of cultural renewal. Wasn't this where the money was? The smart money, at least. New York? Well, wasn't New York suffering from creative malaise? People kept whispering about all the noise coming out of CBGBs (who the hell was Patti Smith? Had you heard of her?), but hadn't the big shiny stuff gone west?

There was no greater signifier than the gentrification of SoHo and the area's transformation from a low-rent district for artists into a playground of uptown chic. In 1965, for a hundred dollars a month, artists could live and work in 2,500 square feet of high-ceilinged lofts in the area south of Houston Street. This was the dawn of loft-living, which over the next ten years would be replicated all over the US, as city dwellers moved out to the suburbs – escaping drugs and violence, and everything that came with them – leaving downtown to be colonised by creatives. For a while, to live in a downtown loft became the goal of any seriously aspiring young artist, empirical evidence that you were not only dedicated to your art, but were successful, too. Having a loft was as important as having a gallery show, often more so. Gallery shows could go wrong, but at the time a loft was there for ever. If a critic destroyed your exhibition, you were going to get over it; if they laid into your apartment then your social standing might take a hit too.

Ten years later, SoHo was no longer cheap, with lofts now selling for as much as $150,000 to doctors, entertainment lawyers and psychoanalysts. Some artists stayed and wrestled with their new neighbours, but others fled, believing SoHo was on its way to becoming another Greenwich Village, another artistic theme park. The artist Paul Harbutt said, 'SoHo can't be isolated just for artists. It's sad. You just can't preserve it as an artists' neighbourhood. The artists have become cynically promoted. They've become technocrats, and the galleries lean to those painters who produce work done with a saleable technique.'

By 1975, while New York itself was on the brink of bankruptcy and the verge of complete chaos, the opposite appeared to be true in SoHo, which in many respects was flourishing, economically at least. For a while it seemed as though art and commerce could cohabit quite happily, but by 1975 the tension felt like it had its very own movement.

Once one of Manhattan's red-light districts – popular with everyone from truck drivers to bankers, from waiters to nightclub tzars – SoHo was now a neighbourhood on the make.

It wasn't just residential properties that were changing, it was retail too, as it became an alternative Madison Avenue. Fashion shops started to open, along with cool bars and chi-chi restaurants. Sure, the area still boasted many from the artistic community – those artists, dancers and musicians who could afford to stay – but the area had flipped. Places like the National Art Workers' Community and the Art Workers' Coalition had vanished, and SoHo was now decidedly apolitical. Robert Indiana was still there, as was Jasper Johns, but younger artists had started moving to NoHo, SoCa and TriBeCa (north of Houston, south of Canal and the triangle below Canal), for the cheap lofts they once went to SoHo for. Saliently, they were also looking at Brooklyn, and of course, Los Angeles, especially Venice and Chinatown. The art critic Judy Beardsalt, who had already abandoned SoHo for TriBeCa, said SoHo was 'changing too rapidly. It's so fluid you can watch it happening.

Soon it will be Greenwich Village, a completely fake Bohemia, filled with dead ideas and dead art.'

The New York Times *said that the soul of the new SoHo was a new publication called* The SoHo News, *an independent newspaper published out of an old, tile-floored ice cream parlour on Spring Street on the western fringe of Little Italy. Its publisher, a former press agent named Mike Goldstein, was determined to take on the* Village Voice, *which was still the left's downtown artistic and political voice piece. 'SoHo is not uptown or downtown,' he said. 'It's a town gone crazy. It's an artistic, creative hustle. SoHo is to the seventies what Haight-Ashbury was to the sixties. Only the location is new.'*

All of this was true, but a greater truth was the fact that SoHo was actually preparing itself for the eighties, when New York would quickly become an island dedicated above all else to commerce. Bohemianism was ripe for lifestyle appropriation, which would be the way in which culture would develop over the next few years, a consumer articulation of cultural transgression, the monetisation of all forms of creativity. The way in which the public would consume any kind of art from now on would be in a shop. That shop could come in any form – a gallery, a nightclub, a dance space, even a restaurant – and in any shape. In this case, that shape was the shape of SoHo, from Canal up to Houston and from West Broadway all the way to Lafayette. Look out, here come the eighties.

The New York Times *couldn't stop writing about the place, comparing SoHo tourists to birdwatchers coming to the area to witness exotic plumage.*

Axe Britannia

This was zenith rock. Mount Rushmore rock. By 1975, Led Zeppelin were music's great colossus, the perennial soundtrack to mayhem and carnage, a swashbuckling band yoked to all manner of imaginary rampaging hordes. They were the hard rock thunderous blitzkrieg, a gang of marauding Viking warriors (albeit in double denim), the template of seventies' orthodoxy and the bar by which every other rock group was then judged. Little did they know it, but they were about to reach their peak.

Physical Graffiti by Led Zeppelin

IT PROMISED TO BE A night dripping in sex and decadence. Between 7–8 p.m. on 31 October 1974, a series of Rolls-Royce Phantoms and stretch Mercedes limousines started to pick up the evening's guests in London. Safely ensconced they got stuck into the cars' well-stocked bars. Vodka, tequila, whatever they had requested, with a recently plenished ice bucket.

Tonight, these several hundred guests were being ferried all the way down to Kent, where the incomparable Led Zeppelin manager Peter Grant had hired Chislehurst Caves, the labyrinth of centuries-old, man-made tunnels near Bromley, for a Halloween extravaganza. The occasion was the official launch party of the first release on the band's new record label, Swan Song, the next stage of the group's journey to global domination. The invitation on the backseat said it all: a large, black, embossed card complete with gothic script and pseudo-hysterical language encouraging the participants to *'do what thou wilt . . . but know by this summons, that on the night of the full moon of 31 October 1974, Led Zeppelin request your presence . . .'*

The night promised the delights of the devil . . . and what a night it turned out to be. On arrival, the first things the guests saw was a phalanx of young men dressed in medieval yeoman outfits holding tiki torches, as though they were guarding the set of a Fellini movie. Once inside the caves, the guests were greeted by naked girls wrestling in open-top coffins which had been filled with jelly. They pretended to bring each other to climax, before starting the whole libidinous process again. Elsewhere, there were waitresses dressed as nuns (with stockings and suspenders, obviously), a troupe of small people who were tumbling and standing on each other's shoulders, fire-eaters, jugglers and George Melly and the Feetwarmers working their way through a set of sexy trad jazz (Melly's signature song was a dirty little ditty called 'Nuts'). The DJ and TV presenter Bob Harris was there that night and he said it was like nothing but a medieval orgy, although one wondered how he knew. The music was deafening, the food never-ending, and everywhere you looked was flesh. As a benchmark of rock-and-roll excess, the launch of Swan Song would be remembered as the most decadent, the most indulgent and the most outrageous party of the decade.

But now that Led Zeppelin had their own record label, what were they going to do with it?

Guitarist Jimmy Page doesn't remember much about the party, but what he does remember is that he'd never been to the caves before and that he found the entire evening quite difficult to navigate. 'It was all decked out with candles and was quite spooky,' he told me. 'Groucho Marx came to the US launch and his way of signing autographs was to trace your hand and sign in "Groucho Marx". How cool is that? There was another party in Covent Garden when someone said they got thrown out of a window by Peter Grant, which I suppose could have happened!'

By the mid-seventies, the double album was perceived as the defining artistic statement, one that had already been made by the Beatles'

'White Album', the Who's *Tommy* (and then *Quadrophenia*), the Rolling Stones' *Exile on Main Street*, and of course Bob Dylan's *Blonde on Blonde*, which had been first, released in June 1966, and Frank Zappa's Mothers of Invention, whose *Freak Out!* had come out a few weeks later. Artists didn't have to curtail their message, edit their songs or think about boundaries to creation in the act of inception at all. This was the great liberating moment for modern music. It also became something of a badge of honour: were you big enough, good enough, or ambitious enough to carry it off? To announce the release of a forthcoming double album let the great unwashed know that you had too many ideas to try and squeeze onto a mere *two* sides. Single albums were for mortals. Those below the salt. Peripherals. Jeez, by 1975, Todd Rundgren had released not one double album, but two!

Surely it was only going to be a matter of time before Led Zeppelin joined their peers in the pantheon of length. For the band, most of 1974 had been spent setting up Swan Song and navigating the chaotic production of their film *The Song Remains the Same* (eventually released in 1976). They had also found time to start recording their next album. In the spring, they retired for a third time to Headley Grange, a Grade II-listed eighteenth-century estate in Hampshire where they had recorded *Led Zeppelin IV.* They had stockpiled enough songs for at least two records and Page elected to try and use as many of them as possible.

The band were now filled with such a huge sense of their own importance that turning their output into a double album was perhaps inevitable. What they ended up with was, if not exactly their defining moment, then at least the apex of their achievements. The purpose of the record – which would be called *Physical Graffiti*, coined by Page to illustrate the energy that had gone into producing it – was to show that they had an embarrassment of riches.

'We took a year out, but in the meantime, Peter Grant had been talking to Ahmet Ertegun [from Atlantic, Zeppelin's record label] about the group having its own label, which became Swan Song,' said Page. 'Bad Company and Maggie Bell had already been signed, but it made sense for our next album to be on Swan Song, a flagship album. I start writing the riffs at home and tracking on them, things like "Ten Years Gone", "Sick Again", "Kashmir", "In My Time of Dying" and "Wanton Song". Then in October '73 we went to Headley Grange to start recording the album. For the first few days there was just me and Robert [Plant] and John Bonham, and we just started playing rock and roll stuff – "Teen Beat" by Sandy Nelson, Presley, rockabilly, just to get together and have fun.'

Page then started going through what he'd recorded at home with drummer John Bonham, as the relationship between Page and Bonham was always the core of the band: the interplay between the guitar and the drums, the sonic and rhythmic bedrock of Zeppelin. 'I wanted every album to have a different personality, so they all sounded different sonically, which is exactly how it worked out. The first album was rehearsed and routine [practised] at my house in Pangbourne. The second album was all about dynamics. The third album showed the acoustic side a bit more, and the fourth album was a mixture between the acoustic and the electric. But it was always important for me to sketch things out with John Bonham, as there was always so much improvisation. We locked in, with a fusion between the guitars and the drums. There was a great understanding between us. We played "Kashmir" over and over, just the two of us, because it was just this great hypnotic beast, with him appearing to reverse the beat. I knew it was going to be like a round, with all these wonderful cascades on top. I couldn't wait to play him the material as I knew he would really enjoy it.'

Page started going through the material the band had left over from their last few albums, which is when a double album started to appear. 'I thought that the first Swan Song album demanded that. I had character pieces, and songs that showed a different side to us. It was varied. They were time capsules. Flasks. But when we got into the groove, it was wonderful. It was a great working unit and there was a real collective consciousness there. *Physical Graffiti* was a fine double album. It has got its own character.'

The album was a quantum shift from what had come before: their 1969 debut, *Led Zeppelin*, which invented the seventies in the space of forty-four minutes and fifty-four seconds, and at a cost of just £1,782 (one of Page's original names for the group was the more prosaic Mad Dogs; they had only been together for two and a half weeks before they recorded); *Led Zeppelin II*, also from 1969, the heaviest rock album ever made; 1970's *Led Zeppelin III*; *Led Zeppelin IV*, from 1971, which contained 'Stairway To Heaven' (unceasingly voted the greatest rock song ever recorded, for a while this became the most played track on US radio. It was so beloved by aspiring guitarists that it was actually banned from being played in some Denmark Street guitar shops in London); and 1973's relatively lacklustre *Houses Of The Holy*.

Physical Graffiti was not only officially their acknowledgement of what would soon become known as world music, it showed a variety that had been previously missing from the group's earlier releases. The group had always been band leader and guitarist Jimmy Page's vision of what he thought a modern rock band should be – explosive, dynamic, all-conquering, the last word in savagery – and yet he had worked with bare-chested singer Robert Plant, bulldozer drummer John Bonham and inevitably quiet bassist John Paul Jones to produce a sound that could never be categorised only as heavy metal. *Physical Graffiti* was more extreme in its diversity. When you listened to early Zeppelin you could imagine the four of them standing tall, standing proud, putting their hands on their hips (perhaps under the mighty brow of a prophetic mountain) and surveying the skyline, almost as though their music was being made without them. In a sense that wouldn't have been so surprising, because as Zeppelin's extraordinary sound started to become so otherworldly – it was on *Led Zeppelin II* that the futuristic brutality of their noise began to take shape – it became easy to assume that this really was the music of the gods, with Page and co acting as mere conduits.

In their time, these conduits certainly attracted their own disciples, because in the first half of the seventies most young men between the ages of fifteen and twenty-five tried to look like Page or Plant:

shoulder-length locks, billowing flares (covered perhaps in one of the band's rune-like symbols), maybe a velvet jacket and a pair of platform boots. It was during the cooler months when their disciples could be mistaken for a real army, however, as they would wander around in old army greatcoats – the type with big, fat belts – possibly holding a Zeppelin album under their arm, to show their allegiance. For some reason – probably because of its extremely recognisable cover, which was based on a photo of Manfred von Richthofen, the 'Red Baron', and his 'Flying Circus' Jagdstaffel 11 squadron during the First World War from 1917 – this was usually a copy of *Led Zeppelin II*. Not only did Jimmy Page's band sound like nothing on earth, they managed to co-opt an entire generation of decidedly earthbound devotees.

The band always felt that too much explanation of their work or the examination of its origins was unnecessary, yet at their heart they were a modern blues band, a heavy one at that. And if you aspired to be a member of the rock fraternity in the early seventies, you were judged on how 'heavy' you were – how loud, how showy, how dynamic. If your power chords were riotous and barbarous and 'authentic' enough (whatever that meant and, actually, no one ever really knew) then you were allowed into the fold. Zeppelin were universally considered to be the heaviest group of them all – Page's riffs and power chords had monumental strength – and so consequently they were often deemed to be the coolest.

The band also became a byword for debauchery and excess. Everything they did was on a grand scale: comestible-covered groupies seemed to be readily available, drummer John Bonham could be seen riding motorcycles down hotel corridors, while rented rooms were regularly trashed and 'redecorated'. Once, when a hotel receptionist said it must feel great to throw a television through a window, the band's legendary manager, Peter Grant, took two hundred dollars out of his wallet to cover a new set and said, 'Here, be our guest.' One story has Page being delivered to a waiting throng of girls on a room service trolley. Their sexual extravagance was mirrored in some of their songs: during 'Communication Breakdown', for instance, Robert Plant can be heard to scream, 'Suck it,' just before Page delivers a ferocious guitar solo. While this seems unconscionable now, it was symptomatic of the age. More menacingly, Page had a fascination for the occult, especially the work of the author and magician Aleister Crowley. This allowed the increasingly copious number of Zeppelin fantasists to paint ever-more colourful narratives of the band's so-called 'deal with the devil'. Of course, none of it was true, but it was great for business.

There were many who thought Zeppelin were pretentious and preposterous and actually rather full of themselves, but the band had the weight of numbers behind them – they were the biggest concert attraction, the greatest album sellers, the loudest! – and for the first half of the seventies were as unimpeachable as Clint Eastwood, Jack Nicholson or the Rolling Stones, outcasts who had been adopted by the mob. They weren't all liked by the same people, but their level of fame was such that it made any kind of qualitative criticism redundant.

Perhaps sensing they were irreproachable, for a while they became rather imperious, which obviously turned the critics against them, critics who understood the mass appeal of the music, but who couldn't fathom the band's disregard for the fourth estate. Thus, they were painted as scoundrels and so became even more hip.

Today it might please pop historians to paint Zeppelin as the prime exponents of an outmoded and justifiably redacted part of the post-war music narrative, but in their time, they actually were the coolest band in the world. In their sonic pomp, Zeppelin were the mothership of motherships. Monstrous, thunderous, epic.

And Jimmy Page was the man who made it all happen. From Heston, Middlesex, Page was born in 1944 and in his youth was obsessed with just two things: art and music, notably the guitar. He largely taught himself by listening to the solos on records by Elvis Presley, British folk artist Bert Jansch and a host of vintage blues guitarists. It's fascinating to learn, however, that a lot of his early influences came from classical music.

'My introduction to the world of recorded sound came when my family lived in Feltham, London,' he told me. 'There was a neighbour on our road who'd recently acquired a top-of-the-range stereo record player and he was inviting neighbours to come and listen to his prize possession. We went round to his house – I would have been around seven years old at the time. He played these audio file recordings for hi-fi enthusiasts, including a steam train, like a Flying Scotsman, zooming across from the right speaker to the left with all its undeniable drama. Curiously enough, that's just the kind of thing I did later with tape recorder facilities when I was playing live with the Yardbirds.

'He also played some stereo classical music on his system, and it was a listening experience that really opened up my ears. At home, we had our little radio with a little speaker, but it couldn't compete with the magnificence of a classical orchestra in our neighbour's house. My parents occasionally listened to BBC radio at home. However, through my neighbour's hi-fi, I actually heard and felt a full orchestra in stereo for the very first time. It was probably something like Elgar or Wagner, a really passionate piece. The whole landscape of music, and the depth and texture of it, really affected me. I don't think I'd ever listened in such detail before.'

In the sixties, Page became the most sought-after and prodigious session player in London, playing on hundreds of records by the likes of the Who, the Kinks, Donovan, Lulu, P. J. Proby, Burt Bacharach and Cliff Richard. In 1964 alone he worked on Marianne Faithfull's 'As Tears Go By', The Nashville Teens' 'Tobacco Road', the Rolling Stones' 'Heart Of Stone', Them's 'Here Comes The Night' and Petula Clark's 'Downtown'. He even contributed to the incidental music on the Beatles' film *A Hard Day's Night*. In 1966 he replaced Eric Clapton in the Yardbirds and when that ran its course, created the New Yardbirds, who almost immediately turned into Led Zeppelin. The vision was all Page's.

'I had a whole sort of audible vision of what I was trying to achieve with it, because having played with the Yardbirds in America, I could really feel what the scene was over there,' he told me, talking about the genesis of Zeppelin. 'And with the advancement of FM radio, which was starting to play album tracks, I could see where it was all going. They weren't yet playing the whole sides of albums, but I thought, They will! They will! I was aiming at the album market, and I figured if you have something that is so interesting that one track leads into the next and it's not just the same sort of vibe on every sort of song, if you are giving variety, and with all the different styles of guitar I played . . . That's what I wanted to showcase on the first album. I wanted everybody to feel that they're going to be able to reinvent themselves and play the best they'd ever played, which is of course what happened. It was the start.'

Recording *Physical Graffiti*, they worked fast, taping the majority of the songs in one or two takes. Some songs dated as far back as the spring of 1970, with the rest written in the weeks leading up to recording. Even though Page had a laser focus, there were bouts of traditional Zeppelin madness. They took farm animals up to the first floor of the hotel and let off flares. Drugs were taken in vast amounts. Everything stopped for several weeks when one of the roadies, a man called Peppy, drove Bonham's new car – a BMW 3.0 CSl – into a wall. Bonham was so upset, he started telling anyone who would listen than he wanted to kill Peppy, who hid in a wardrobe for thirty-six hours.

'It was just young blokes having a laugh,' said one of the engineers. 'The band had this belief about them then that they were untouchable – as we all do. It was all to do with testosterone and, believe me, Robert Plant had more it it than anybody I've ever known.'

One morning, Bonham arrived for work with a suspiciously large bag containing 1,500 pills of the sedative Mandrax (known in the US as Quaalude), intending to conceal them from the rest of the band by taping them to the inside of his drumheads. A member of the crew spotted the flaw in his plan, pointing out to Bonham that he had a Perspex kit. The drummer was so chemically deranged he would regularly punch his drum roadie in the face for no reason. He would also disappear to the local pub and offer to buy everyone drinks, all day, and all night. Another assistant had to ensure a ready supply of nappies for when the alcohol caused Bonham to lose control of his bladder. Page, meanwhile, was dabbling rather too much in powders and deepening his interest in what he referred to as 'my studies of mysticism and Eastern and Western traditions of magick and tantra'. This was seventies' excess in excelsis.

The recording sessions began with bassist John Paul Jones apparently wanting to quit. Apocryphally, he told Zeppelin's manager, Peter Grant, that he'd had enough of recording and touring and intended to retire from rock and roll to work with the choir at Winchester Cathedral, before eventually being persuaded back.

Physical Graffiti was released on 24 February 1975, in an expensive die-cut sleeve that pictured a New York City brownstone tenement block, photographed in the East Village, through the windows of which you could find Elizabeth Taylor, Lee Harvey Oswald and Robert Plant dressed in drag. It was the first album ever to go platinum on advance orders alone and, for a while, it was the fastest-selling record in history. One Atlantic Records bigwig said, 'The audience was ahead of the company. I never saw an album sell as much. You'd go to stores and there were lines, and everybody was waiting to buy the same record.' For now, at least, there was still no stopping Led Zeppelin. The record was extraordinary, sounding idiosyncratic as well as enormous. At this stage in their career, it wouldn't have been surprising if they had started to sound generic, and yet *Physical Graffiti* was almost a gear change.

Fundamentally, it was ambitious. One reviewer said this was the sound of a group writing their identity, in huge block capitals of sound, across popular culture. In a way, that was true, but it was always a lot more than that. For Led Zeppelin it was a mountaintop moment.

Page was extremely particular – some might say obsessive – about the sonic quality of Led Zeppelin. 'Like Jeff Beck said, I always had a very eclectic record collection, with Indian music, Arabic music, classical, electronic, lots of avant-garde stuff,' said Page. 'I was constantly trying to play various styles at the start, but not necessarily very successfully. I mean, I wasn't as fluent a jazz guitarist as I would have liked, but I just somehow developed my own sort of style. It's all about the player. If you have a guitar and an amplifier and Jimi Hendrix comes into play, he's going to sound like Hendrix and it would be the same with Jeff or Eric.'

As the architect of the whole Zeppelin ethos – the sound, the look, everything – how important in the early days was it that the band adhered to his vision?

'I think one of the things that was really key was that I'd had a lot of experience, both on my own and with bands, especially with the Yardbirds. Robert and John Bonham were slightly younger than us. But because we had the time to rehearse, we quickly developed how the band should sound. We also did a little tour, which was something the Yardbirds would have done if they hadn't broken up. So we could go out and showcase what we wanted to record. It gave us a lot of confidence to go in on that first set of recording sessions, because it was already tried and tested.'

Every Zeppelin album is very different, and he obviously knew exactly what he wanted them to sound like before he recorded them.

'I really did know what I was trying to do with each album and the important thing was making sure they were different from the one that preceded it. We had made a policy not to release singles and so we never worried about recording one.'

Page was a great fan of the virtuosity of the early seventies, the experimentation of Yes, the showmanship of Emerson, Lake & Palmer. For him, it was all a broad church, and one to be embraced. 'Quite early on I got nervous about listening to other things, because I didn't want to appear to be influenced by anyone else. Of course, I heard what other people were up to, but I never let it influence me. I was just more interested in recording at home and experimenting on my own, rather than worrying that I was being influenced by Eric Clapton. I knew there was a lot of criticism of us in the press, but we were too busy working. All we cared about was the quality of what we were doing. Is it any good? That's all we cared about. One thing for sure is that I never worried about how we were going to reproduce something on stage, otherwise there would have been a lot of things I wouldn't have attempted in the first place.'

How important is legacy to him? 'Somebody once asked me if I was intimidated by my past and I wasn't being flash when I said this, but it was actually an immediate response. I said, "No, it inspired me."'

Zeppelin's high-concept, high-octane mix of light and shade, of push and pull and loud and quiet – was nigh on impossible to top. Nineteen-seventy-five was the last time they were inviolate, immaculate and beyond reproach. They remain an incubator of heroic fantasies and it is now impossible to listen to the likes of 'Trampled Under Foot', 'Kashmir', 'Babe I'm Gonna Leave You' or any of their other Wagnerian classics, complete with their wailing and their titanic rock riffs, without imagining yourself as the invading conqueror of something or other – even if you're just overtaking someone on the M40.

Three months after the release of *Physical Graffiti*, the band made a move on Earls Court. For five nights in May, and to a total audience of 85,000, the band confirmed their status as the biggest and loudest band in the country. While the Who rivalled them for excitement, Zeppelin were the relatively newer proposition and, unlike the Who, they wallowed in the pomp. Everything about the Earls Court shows was big, from the gigantic video backdrops to the towering sound system. They played each night for three and a half hours, and they could have easily played for longer had they been allowed to. Fans had queued for twenty-four hours for tickets. *Melody Maker*'s Michael Oldfield said these shows were 'the definitive rock performance', drawing on all parts of the band's seven-year career, as they effortlessly moved from blues-rock to post-hippie pastorals and back again. They played everything they were expected to, with 'Kashmir' becoming the residency's highpoint, night after night.

The most important British music festival of the year was at Reading, in August. Zeppelin didn't appear (they were far too big for that), although Yes did, who were probably the second-biggest band in the country. Over the three-day weekend you could also see Hawkwind (complete with their libidinous dancer Stacia), Wishbone Ash, Robin Trower, U.F.O. and Richard and Linda Thompson alongside Supertramp, Soft Machine, the Heavy Metal Kids, Thin Lizzy, the Kursaal Flyers,

Joan Armatrading and the UK's hottest live act. Dr. Feelgood. This was a pretty representative assortment of British acts at the time, although many people had bought tickets for Sunday night's special attraction, Lou Reed, who was a no show. The threat of his appearance was thrilling in itself and showed the chasm between the appeal of UK and US acts. The first UK appearance of the Ozark Mountain Daredevils might have impressed a handful of British music journalists, but they certainly didn't set Berkshire alight that weekend. Reading was always a B+ kind of experience anyway, and the organisers were unlikely to secure the likes of David Bowie, the Rolling Stones, Roxy Music or indeed Zeppelin themselves; it was a reminder that, with a few obvious exceptions, at the time American music in many ways seemed so much more exotic.

The fact Zeppelin didn't appear – and would never appear – set them apart from the pack. By the time they played Earls Court, Zeppelin had already played thirty-six dates in the US. They were now spending more time and money accessorising the sets than on previous tours, following in the path of the Rolling Stones, for whom arena shows had become state-of-the-art experiences. Bonham's drums would be on a riser, there was a fancy new light show, dry ice, lasers, neon and the band were even learning to dress up, Plant wearing a silken, kimono-like wrap over his tight jeans, Page sporting expensive, beautifully embroidered stage suits, and Bonham wearing a white boiler suit and a bowler hat, the menacing garb of the droogs in *A Clockwork Orange*. In a further nod to the film, Plant began introducing Bonham as 'Mr Ultra Violence' and occasionally as 'Mr Quaalude, the man who set fire to his own bed!' Bonham was 'our protean percussionist – the Lord of the Droogs – a man you never want to meet in a dark alley.' This was when Plant wasn't calling Bonham 'Karen Carpenter', a reference to the fact she had been voted a better drummer in *Rolling Stone*.

As the tour moved across North America the band as a unit became more confident and therefore more versatile. On Long Island on Valentine's Day, Plant told the crowd they would be playing all the flavours of 'the glorious ice-cream cake of Led Zeppelin'. The band were also leasing the *Starship*, a converted United Airlines Boeing 720 passenger jet, bought by singer and actor Bobby Sherman and his manager Ward Sylvester, which Zeppelin had used once before on their 1973 tour. A gigantic runic Led Zeppelin logo was painted on the fuselage of what was described by many as a flying gin palace. The *Starship* cost $2,500 per flight hour to lease, meaning a typical cross-country trip would cost whoever was renting it around twelve thousand dollars. The plane's organ had been played by everyone from John Paul Jones and Elton John to Stevie Wonder and John Lennon.

Elsewhere in the US in '75, Bob Dylan played the first twenty-two dates of his comeback Rolling Thunder Review, Bruce Springsteen played eighty-two theatres from Asbury Park to Buffalo, and Fleetwood Mac played ninety-five gigs from El Paso to San Diego, the first tour with new recruits (and saviours) Lindsey Buckingham and Stevie Nicks. The Who, Queen, Wings and Genesis were all also on the road, although the most impressive schedule was orchestrated by Kiss – they were constantly on the road this year, playing 126 dates in the US, as part of three tours which ran into each other. It was almost as though they were playing a cross-country residency.

Just before the New York Dolls appeared on *The Old Grey Whistle Test* in November 1973, the Dolls frontman David Johansen told the show's presenter Bob Harris he had 'bunny teeth'; following the Dolls' vigorous and super-camp mime through their song 'Jet Boy', Harris sniggered into the camera and pronounced the band 'mock rock'. Far from sentencing them to a career of dismissal and oblivion, the programme's audience seemed to enjoy the Dolls. They were, like Alice Cooper or the Sensational Alex Harvey Band, an example of how theatrical parts of the rock fraternity had become. Cooper in particular had managed to turn his stage persona (he was christened Vincent Furnier) into a commercial success. By '75 he had binned his band and become a solo act, emphasising the ghoulish nature of his performance, stalking the stage drenched in the macabre and covered in make-up. On 17 June he played in front of seventeen thousand people at the Forum in Inglewood (supported by Suzi Quatro), and all hell broke loose. This was his Welcome To My Nightmare tour, which had been advertised as a proper horror hoot. On a screen a haunted graveyard was projected, as dry ice swept over the crowd and Cooper muscled his way through various songs from the new record, including the hit 'Only Women Bleed', along with a selection of greatest hits – 'No More Mr. Nice Guy,' 'Billion Dollar Babies' and 'School's Out'. The denouement was always the same, with Alice guillotined and his body being put in a coffin before being hauled off by his pallbearers. As Alice said himself, 'We were the group that drove a stake through the heart of the Love Generation.'

Pink Floyd were born of the Love Generation, but by 1975 they were a very different proposition altogether, a space-age behemoth whose music was grandiose and self-referential. With Roger Waters's fearful, dystopian lyrics, and David Gilmour's almost architectural guitar, Floyd's records suggested the intergalactic rather than the whimsically psychedelic. For five nights in April, they premiered the new material from *Wish You Were Here*, the follow-up to their monumentally successful opus, *The Dark Side of the Moon*, at the Sports Arena in LA (capacity 16,000).

Zeppelin were bigger than anyone. In the US, the band maintained a reputation as outlaws of the road, while talk of their excesses in hotel rooms ranged far and wide. 'Like the music, the legend grows too,' said Robert Plant, when he was asked about their behaviour. 'There are times when people need outlets. We don't rehearse them and, let's face it, everybody's the same. No, we're not calming down yet. Calming down doesn't exist until you're dead. You just do whatever you want to do when you want to do it, provided there's no nastiness involved. Then the karma isn't so good.' Zeppelin would eventually succumb to karma, but the legend of *Physical Graffiti* would only grow and grow.

All Aboard for Chi-town

3 June, 46th Street Theatre, Broadway, New York

YOU MIGHT HAVE THOUGHT THE city had been venerated enough, at least in song. Frank Sinatra famously had two stabs at it and could have easily had more. The first was 'Chicago (That Toddlin' Town)', written in 1922, when the town was roaring (the original third verse included the lines, 'More coloured people up in State Street you can see, Than you'll see in Louisiana or Tennessee') and recorded with Nelson Riddle in 1957. He recorded the second, 'My Kind of Town', written by Jimmy Van Heusen and Sammy Cahn, in 1964, originally for the Rat Pack movie Robin and the 7 Hoods *(nominated for an Oscar, it lost to 'Chim Chim Cher-ee' from* Mary Poppins*).*

But John Kander and Fred Ebb's Chicago, *based on a 1926 play by Maurine Dallas Watkins, would not only become synonymous with the city, but would also eventually become Broadway's longest-running American musical. Set in twenties' Chicago during the jazz age, and based on real-life murders and trials, the plot concerns two jazz-age murderers, Roxie Hart and Velma Kelly, who compete for the attention of the Chicago tabloids and become overnight media sensations. Watkins based Roxie and Velma on two criminals she had covered for the* Chicago Tribune: *Beulah Annan, whom she dubbed 'the beauty of the cell block', and Belva Gaertner, the 'most stylish of Murderess' Row'. Roxie and Velma vamped through the show like Liza Minnelli did in the film version of Kander and Ebb's previous hit,* Cabaret *(few musicals offered equal-value roles for two stars of the same gender, but this was one).*

With its louche, razzle-dazzle choreography by Bob Fosse, Chicago *took a jaundiced view of the nexus of crime, the media and celebrity. It was also intentionally kitschy. And audiences loved it. The show's hot, jazzy music was the real star, as there was nothing vague or glancing about the songs. They were hits, 'All That Jazz', 'Cell Block Tango' and 'When You're Good to Mama', in particular. Kander's music, as exhilarating as a bottle of the very best bubbly, and the chorus line dressed in black, revved the engines from zero to sixty in a show that demanded you hopped aboard or got left behind.*

The show would go on to define America time and time again. In 2017, the Washington Post's *Nelson Pressley said, 'What's clear is that the laughing spirit and corrupt heart captured by the jaunty vaudeville songs really is America, now and always. Is truth getting trammelled by spin? Buzz? Misdirection? Is some shameless flimflam artist riding high? When this revival opened in 1996, comparisons were quickly drawn to the 1995 O. J. Simpson trial, with a celebrity skating away as the term "Kardashian" entered the lexicon.' In* Chicago, *as in real life, someone is always getting away with murder. Pressley pointed out that this sexy, seedy celeb-fest felt like a national monument, a ravishing Grand Canyon where ethics go to die and American Dream opportunism prances and gloats.*

Sinatra was almost conjoined with the city of Chicago, and not least because of his mob connections. He was a leviathan, a man who prowled the night looking for wine, women and an amplified microphone, whether he was in Nevada, New York or Chi-town. He only lived once – and, the way he lived it, once was enough. In Frank's world there was always a silver lining if you knew where to look for it. He also had strict rules, maxims you ignored at your peril: 'You treat a lady like a dame, and a dame like a lady.' 'Alcohol may be man's worst enemy, but the Bible says love your enemy.' 'Cock your hat – angles are attitudes.'

And what amber-drenched nights he had: after-hours marathons when he'd drink nothing but tumbler after tumbler of J. D. He once went to see a new doctor who asked about his alcohol intake. Sinatra, somewhat taken aback, said he drank thirty-six a day. The doctor, perhaps unsurprisingly, asked him to be serious. Sinatra said he was serious, and that he drank a bottle of Jack Daniels every day, which he reckoned was roughly thirty-six drinks. Appalled, the doctor asked him how he felt when he woke up every morning. Frank said, 'I don't know. I'm never up in the morning and I'm not sure you're the doctor for me.'

He was still performing in 1975, still racking up the concert dates as though he was never going to stop. Terry O'Neill photographed him in London that year, singing at the Palladium with Count Basie and Sarah Vaughan. He had been photographing Sinatra for years, after an introduction from his friend Ava Gardner. While Sinatra was shooting the Tony Rome film Lady in Cement *in Miami,* Life *magazine sent Terry down to the set to shoot him. Unhelpfully, but not unusually, no access had been sought or agreed, and so Terry had to use his powers of persuasion to convince Sinatra to collaborate. He asked Gardner if she would write a letter of introduction to her ex-husband, which she duly did. Days later, sitting on a boardwalk sun lounger, waiting for Sinatra to turn up, he saw him approaching, accompanied by a phalanx of bodyguards. 'I saw him walking round the corner and I just thought, "Click!"' said O'Neill. 'As soon as I'd taken the picture, I thought, my God! What have I gotten into here? I gave Frank the letter. He read it, smiled, turned to his guys and said, "The kid's with me."' And suddenly he was.*

He stayed with him for three weeks. 'It's a cliché to say that photographers hide behind their camera but that's absolutely true of me,' said O'Neill. 'Sinatra taught me that, to hold back. I spent such a long time with him, went everywhere with him, but we barely spoke. And then I realised that was the secret: being allowed in the inner circle, but not being afraid of them. That's how I got such candid shots. That's how I took pictures. If you could survive him, you could survive anyone, because one mistake and you were out.'

He Wore Electric Boots
and Mohair Suits

If the seventies delivered anything, it delivered fame. Being famous in the seventies meant more than being famous in any other decade. The twentieth century may have already delivered the likes of Valentino, J. F. K., Sinatra, Marilyn, Elvis and the Beatles, but previously, fame was sporadic. By 1975 it was industrial. At the time, Elton John was probably the most popular entertainer in the world. The Eagles may have sold more records, but they weren't famous like Elton was famous. In 1975, nobody was.

Captain Fantastic & the Brown Dirt Cowboy by Elton John

ELTON JOHN WAS BACKSTAGE AT Dodger Stadium, and he seemed satisfied with what he saw in the dressing room mirror: a shimmering Bob Mackie-designed sequined Dodgers baseball uniform and cap, with 'ELTON' and the number '1' in Dodger blue on the back. He was wearing contrasting socks, Nike trainers, oversized white glasses and he was sparkling. In fact, he looked like the most sparkling thing in all Los Angeles. He felt sparkling, too.

Los Angeles in the seventies is forever associated with two images. The first features Faye Dunaway reading the papers, poolside at the Beverly Hills Hotel, the morning after winning the Oscar for her performance in *Network* at the Academy Awards. The photograph was taken by Terry O'Neill (who would later marry Dunaway), who also took the second image – Elton John on the blue-carpeted stage at Dodger Stadium in October 1975. Elton is leaning against his piano (also covered in blue carpet), wearing the Dodgers kit and is about to launch into 'Benny and the Jets' in front of the 80,000-strong crowd. The picture is so vivid you almost expect it to start playing the song, like a musical birthday card.

O'Neill first met Dunaway in 1970 on the set of a long-forgotten western called *Doc*, a film that actually includes the line 'One hand of five-card stud; my horse against your woman'. He met her again when he was hired by *People* magazine to shoot her for a special issue devoted to the Academy Awards, an assignment that resulted in not just one of his better-known photographs, but also in one of the most famous Hollywood portraits of the decade. Bursting with ennui, this bittersweet picture shows a languorous Dunaway, poignantly alone with her golden statue beside the chaise-lined pool, wearing a peach satin robe and a faraway look, the ground around her high-heeled slippers strewn with newspapers full of celebratory award-night reviews. Ironically, having just taken one of the most important pictures of his career, O'Neill put this part of his life on hold while he tended to something more important. He had fallen in love.

At this point Dunaway was one of the top five box office draws in the States, an iconic beauty whose love affairs – Lenny Bruce, Steve McQueen, Marcello Mastroianni – were as high-profile as her movies. Her relationship with O'Neill was no different, only more protracted, and they soon became an established Hollywood couple, roaming the Spago bistro and the Polo Lounge like a latter-day Liz and Dick. They eventually married, adopted a son, and stayed together until the mid-eighties, living in California, New York, Connecticut, Long Island, and, during a much-publicised sojourn in London, in a little house just behind Harrods.

'The Oscar picture was like trying to compose a painting,' O'Neill said. 'I had to rehearse it rather more than people think, and it took some doing to convince Faye to do it. It worked in the end because the elements conspired to send a powerful message.'

While it is unlikely that Terry O'Neill would have ever been gauche enough to say he discovered Elton, the photographer was one of the first people outside the music industry to take what you would call a keen interest in him. As the sixties turned into the seventies, and having spent a decade cataloguing the fancies and foibles of the Hollywood A-team, O'Neill found himself seeking other trophies. He moved into sport, and with more diligence, music. And Elton was the first person to seriously catch his eye.

It was almost a patronage, as O'Neill would follow him everywhere, capturing his every move for years on end, as he rose from crude, bar-room vamper to stadium superstar. By 1975, he had taken even more photographs of him than even his great friend, Michael Caine. Unlike the acting fraternity, musicians tended to be gregarious creatures – particularly socially disposed gadabouts like Elton – and consequently many of O'Neill's portraits of the portly troubadour showed him fraternising with other pillars of the celebrity circuit: Cary Grant, Martina Navratilova, Candice Bergen and Liz Taylor. To him, the famous were little more than moths.

'I spent so much time with him I eventually had carte blanche to photograph what I liked,' O'Neill told me. 'Of course he was capable of mood swings, but I don't think I ever met a performer who takes so much delight in pleasing an audience, whether he's playing in front of fifty thousand people or showing off to half a dozen in his dressing room. When he did the two concerts in Dodger Stadium in LA in 1975, it was considered by everyone as the pinnacle of his career, yet I found out afterwards that he'd apparently only recently tried to commit suicide. He was up and down like a yo-yo.'

'I had had my stomach pumped, and then two days later I was onstage,' Elton told me once in LA, coincidentally in the Polo Lounge at the Beverly Hills Hotel. 'I've got a very strong constitution. I wasn't going to let a misguided suicide attempt stop me from performing at Dodger Stadium.'

O'Neill embroiled himself in Elton's world. Some of his pictures were little but exercises in iconography – as if to confirm what Freud said were an artist's primary motivations: namely fame, money and beautiful lovers – while others simply reflected the photographer's ability to further the art of reportage. With Elton, both were true. O'Neill spent an inordinate amount of time setting up the Elton photo, and he spent the first night working out where he was going to shoot on the second. He liked the colours in his frame. He said it all looked like a Hockney painting. 'In October 1975, no one was bigger than Elton John. He was like Elvis at the height of his career,' said O'Neill. 'Elton still is one of the most talented people I've ever met, and he gave his all at those concerts.'

Elton remembers one moment in particular from the two Dodger Stadium shows. 'I believe it was the second day, and we were singing "Don't Let the Sun Go Down on Me". The sun was going down behind Dodger Stadium and we were onstage watching it go down, and everyone in the crowd had their Zippo lighters out. You could feel the hairs go up on everyone's necks. I've done a lot of concerts in fifty years. That was the most profound moment I've ever had onstage.' Terry remembers that 'Elton cried after the concert was over. I loved him for that'.

At the time Elton was the biggest star in the world, and his two shows at the home of the LA Dodgers that year were the pinnacle of his early success. A few days before the shows, a chartered Boeing 707 brought Elton's mother, stepfather, grandmother, record label staff, journalists, and friends so they could participate in the celebration. They took tours of Universal Studios and visited Disneyland. His manager John Reid even hosted a party on his yacht. The mayor of Los Angeles, Tom Bradley, declared it 'Elton John Week'. A star was unveiled on Hollywood Boulevard, where Elton arrived in a gold golf cart with a giant pair of illuminated glasses and a bow tie on the front. There were so many onlookers that Hollywood Boulevard had to be closed. And his two Dodger Stadium gigs were bonkers. Emmylou Harris and Joe Walsh opened the shows. Cary Grant was a backstage guest. The Southern California Community Choir performed; even tennis star Billie Jean King sang background vocals.

O'Neill's picture summed up the heights he had scaled. As the DJ Paul Gambaccini once said, no single photograph better demonstrates the hold a rock star can have over the public. 'Benny and the Jets' was also a quintessential LA record, and you could guarantee you'd hear it on the radio whenever you visited the city. You'd also hear every other great landscape driving song – America's 'Ventura Highway', the Doobie Brothers' 'Long Train Runnin'', and of course Elton's 'Tiny Dancer' – come hurtling out of the rental car's speakers, accompanied, of course, by a flash of neon light, a plume of purple smoke, and a wash of dry ice. In 1975, there were more cars in California than people in any of the other states of the United States, while the Los Angeles freeway system handled upwards of five million cars on a daily basis. The lucky residents of LA County spent an estimated two days of each year stuck in traffic. Visitors could get despondent, but then they'd find themselves driving through the hooded hills of Bel Air, past the mansions and the gate lodges of Beverly Hills, past the exotic Chandleresque haciendas, rustling palms, lawn sprinklers and chirruping crickets, and everything would feel right with the world.

Los Angeles was always close to Elton's heart, as it was here, in 1970, that he became famous. In several landmark gigs at the city's notorious Troubadour in August that year, the piano man became a star in the space of forty-eight hours. It was a true overnight sensation, one that he would never recover from. First came acclaim, then commercial success, and sooner than he could ever have envisioned, global cultural supremacy. 'I was doing gigs in England, and then we started getting good reviews in the *NME* and the *Melody Maker*,' says Elton. 'We were the new darlings. But we were small time. We had just played a Sérgio Mendes gig in Paris and were booed offstage. And while the French promoter was saying I was going to be a disaster, my agent, Joe Paratio said, "He's going to be a big star, this guy. I'm going to book him in America." And then he booked me in at the Troubadour. That night there was magic in the air. There was so much adrenaline. I thought to myself, I've got to fucking do this, I've got to pull this off.'

The whole band were nervous, but as they were such a close-knit unit family, almost, it worked. Nigel Olsson on drums, Dee Murray on bass and Elton. A trio of whirling dervishes, exploding, cartoon-like, on the Troubadour stage. Elton says he took to LA like a duck to water, reimagining it as *The Beverly Hillbillies*, a fictional land producing simply extraordinary, real-time music. During those early trips to LA, he met everyone from Mae West and Van Dyke Parks to Leon Russell and Quincy Jones, from Neil Diamond to Brian Wilson. 'Brian was crazy. The living room was full of sand. He opened the door and went, "Elton John, I hope you don't mind, I hope you don't mind."' Up until this point in his life, Elton's excesses revolved around Merrydown Cider and little else, so he was bamboozled by Wilson's behaviour, especially as he later tried to sell him his piano. When he first played the Troubadour, Elton was wearing a T-shirt, a jumpsuit and flying boots with wings on, all from the boutique Mr Freedom, but by 1975 he was sparkling all over. He had also invented a persona for himself. He took his inspiration from Jerry Lee Lewis and Little Richard, people who did things with the piano that he wanted to do. He knew he couldn't do what Jimi Hendrix could do with his guitar, but he also knew he could stand up and jump on his piano, treating it like a trampoline. He did somersaults and handstands, and no one was expecting that. Certainly no one at the Troubadour. 'I wanted to have fun and I wanted to put on a theatrical display. We just didn't want to stand there. If you went and saw the Eagles, you might as well have stayed at home.'

Los Angeles became a second home, where every day conjured another anecdote. 'I went to see Eartha Kitt sing at a gay club there one night, called Studio One. Before she went on stage that night, she came up to me and said, "I've never liked anything you did except the Transylvania thing," by which she meant "Philadelphia Freedom". I mean, what do you say to that? I remember the agent Sue Mengers invited us for lunch and she rang a friend of hers to ask, "Is there anything they don't eat except pussy?"'

Nineteen-seventy-five was also the year of *Captain Fantastic & the Brown Dirt Cowboy*, one of Elton's classic seventies records. Released in May by MCA Records in America and DJM in the UK, it was an instant commercial success. The album was certified gold before its release and reached No. 1 in its first week on the US Billboard 200, the first album to achieve both honours. It sold 1.4 million copies within four days of release and stayed in the top position for seven weeks. Though they would all appear on later albums, this was the last seventies' release with the original line-up of the Elton John Band (guitarist Davey Johnstone, bassist Dee Murray, and drummer Nigel Olsson).

Captain Fantastic is a concept album that charts the early struggles of Elton (Captain Fantastic) and lyricist Bernie Taupin (the Brown Dirt Cowboy). Elton said, 'I've always thought that *Captain Fantastic* was probably my finest album because it wasn't commercial in any way. We did have songs such as "Someone Saved My Life Tonight", which is one of the best songs that Bernie and I have ever written together. *Captain Fantastic* was written from start to finish in running order, as a kind of story about coming to terms with failure – or trying desperately not to be one. We lived that story.' The intricate cover art was designed by the pop artist Alan Aldridge, drawing fantastic imagery from the Renaissance painting *The Garden of Earthly Delights* by Hieronymus Bosch. It was Aldridge who also designed a famous Beatles jigsaw.

Ninety-seventy-five managed to bury Elton's second album that year. *Rock of the Westies*, which came out in October, was easily the equal of *Captain Fantastic*, and is the great lost Elton John record. An alternative *Goodbye Yellow Brick Road*, it is an uneven but fascinating album containing half a dozen classic songs: 'I Feel Like a Bullet (In the Gun of Robert Ford)', 'Dan Dare (Pilot of the Future)', 'Feed Me', 'Street Kids' and 'Grow Some Funk of Your Own.' The album was introduced by the bouncy lead single, 'Island Girl', which had come out a month earlier. The Caribbean flavour to the lyrics was mirrored by the instrumentation, with Ray Cooper on marimba as well as conga and tambourine. The song was a widespread winner, with Top 20 placings in the UK and Australia, but really rang the bell in the US, where it went to No. 1 in just its fourth week on the Hot 100.

'Before, I just used to write melodies to Bernie's experiences and fantasies,' Elton told *Melody Maker*. 'I identify with this album so much more than anything else I've done. For me it will always be my favourite album. But that's from a purely selfish point of view. Whether it will stand the test of time, who knows? You can only tell in retrospect.'

It is an album that immediately transports you to the Los Angeles of the mid-seventies, with the midday sun and the palms above you, driving through a bright blue Hockney dreamscape, surrounded by Cinemascope billboards for *Shampoo*, *Tommy* and *One Flew Over the Cuckoo's Nest*. There is a cool wind in your hair, the warm smell of colitas rising up through the air . . . All of a sudden, your trouser bottoms get a little wider, your lapels turn into aircraft carriers, your cologne becomes a little more pronounced, your shoes start sprouting three-inch stack heels, and your denim waistcoat is suddenly made of silver lamé. Oh, and guess what? You are now probably sporting a pair of tinted spectacles the size of Texas, there'll be a copy of *Rolling Stone* on the passenger seat, along with a packet of More cigarettes, a paperback of Robert Pirsig's *Zen and the Art of Motorcycle Maintenance*, and an eight-track cartridge of Supertramp's *Crisis? What Crisis?*

Nineteen-seventy-five was the year he bought Woodside, his sumptuous thirty-acre estate near Datchet in Old Windsor. With eight bedrooms, five reception rooms, a billiards room, squash court, swimming pool and video and book libraries, this Queen Anne-style manor house felt almost royal. Being 1975, it was also extravagantly decorated. At the time, Elton was taking cocaine on average every four minutes, and as he was personally involved in the decoration and furnishing of his new quarters, it is perhaps not surprising that its execution was so, er, particular. The mansion sat on the edge of Windsor Great Park, originally built for the surgeon of King Henry VIII in the 1500s. The house needed a view of Windsor Castle so that if an emergency flag was flown, the surgeon would know to go to the king's side. The grand manor had been destroyed and rebuilt several times in the centuries since it was built, and when Elton first moved in, he quickly made it his own with a number of kitsch decorations, including a replica of Tutankhamun's state throne. Few were surprised. The Italian fashion designer Gianni Versace used to say, 'Less is less, and more is more,' while his friend Elton would add, 'And more is good. A lot more is very good.'

That June, Elton was due to play a concert at Wembley Stadium, supported – we soon found out – by the Beach Boys, Joe Walsh, the Eagles, Rufus and Stackridge. The DJ and compere was going to be Johnnie Walker. For weeks at school this was all we could talk about, as half-a-dozen of us were going to go. We planned how we were going to get there, what we were going to wear, and what we were going to drink en route. In the end, it all fell apart – how could it not? We were fifteen! – but we soon found out we'd dodged a bullet. The Beach Boys had played a set consisting largely of their greatest hits, so Elton's decision to play *Captain Fantastic* in its entirety didn't go down well with the audience, especially as so few of them had heard it. Which is why so many that night left the stadium early, wandering off to a mythical West Coast idyll, via Wembley Park Stadium, in search of cold cocktails, a welcoming neon bar, and a hardwood floor well-used to the agglomeration of imported cowboy boots.

We're Going to Need a Bigger Movie

20 June, in 465 movie theatres from Hollywood to Palm Beach

JAWS *WAS THE MOVIE THAT ate Hollywood, chewing its weary old bones as though it were a floating carcass dumped in the ocean by a passing trawler. After 1975, an industry that was barely sixty years old would be changed for ever. The movie business had withstood the advent of television and rock and roll, as well as a new generation of consumers who were less impressed with what it had to offer than their predecessors, but it had no chance against the devouring mass of a three-tonne shark.*

Hollywood had started to change after the success of George Lucas's American Graffiti *in 1973, which was deliberately targeted at a younger demographic than was usual, even for a teenage movie (which were treated as inconsequential, specialist films). For many in the industry this was a lightbulb moment.* Jaws *was released at the beginning of August and became a hit largely because of teenagers enjoying school holidays. The producers saw an opportunity to take advantage of this new trend and devised one of the first television marketing campaigns in cinema history. Universal Studios spent a whopping $1.8 million, scheduling a record number of trailers on major TV networks to pique the public's interest in the days leading up to the release. The response was unprecedented.* Jaws, *which cost roughly $9 million to make, would go on to gross $260 million in the US alone, breaking records set by previous blockbusters like 1972's* The Godfather *and 1973's* The Exorcist *and the equivalent of some $1.5 billion today. The industry had never known anything like it. It was released in 465 theatres nationwide*

155

– unheard of for the time – and became a phenomenon, helping create the summer movie season as it stands today, fifty years later. After Jaws, *studios would begin to market movies differently, creating blockbusters specifically for the summer market.*

Based on the Peter Benchley novel, the premise was simple: when a killer shark unleashes chaos on a beach community off Cape Cod, it's up to a local sheriff, a marine biologist and an old seafarer to hunt the beast down. And that was it. Critics were as dazzled as the moviegoers, finding a childlike exhilaration in Steven Spielberg's pacing and direction. 'It's a noisy, busy movie that has less on its mind than any child on a beach might have,' wrote Vincent Canby in the New York Times. *'It has been cleverly directed by Steven Spielberg (*Sugarland Express*) for maximum shock impact and short-term suspense and the special effects are so good that even the mechanical sharks are as convincing as the people.'*

It was a phenomenon, so it soon became something that everyone needed to have an opinion about; consequently, everyone trooped off to see it. They went to see Jaws *because they had been conditioned to want to see it, and then invariably liked it. Loved it, in fact. After all, if you loved it then it was much easier to talk about it, especially as everyone else that year appeared to like it too. Spielberg would immediately be accused of infantilising his audience, which is certainly true, although he created that audience too. A new generation of moviegoers who were intrigued by the European indulgences of the Robert Altmans, Martin Scorseses and Peter Bogdanovichs of this world – who were determinedly exploring the concepts of adulthood – were also prepared to lose themselves in a contemporary version of old-time escapism. More than prepared, actually. It was a wonderful rollercoaster of a movie, and one that felt terribly old-fashioned while still feeling contemporary. It had the added bonus of making younger viewers feel sophisticated, while making older people feel young.* Jaws *was a commodity if it was anything, a fabulous prize at the end of the working week, like renting a sexy red convertible for the weekend. The irony of course was the fact that this wonderful, summery, beach-towel of a movie was all about people being attacked by a giant shark.*

Was it simplistic? In some respects, yes, although it was also a slice of feelgood designed to comfort those who had been brutalised by Vietnam and the vagaries of Watergate. We didn't know it yet, but Spielberg's later movies would obsess over the family, celebrating the fifties' ideal as a way of manufacturing a better family for himself. If most of his contemporaries were rushing as fast as possible away from their pasts, Spielberg was rushing back towards his, reinventing it on the way. There was an element of postmodernism about his films, but never irony. Not ever.

The production had a great backstory, too, as the three principal actors – Richard Dreyfuss, Robert Shaw and Roy Scheider – found it incredibly hard working together. Impossible, really. Jaws *was notoriously difficult to make. It went way over budget, the production dragged behind schedule and the mechanical sharks they were using frequently broke down. The delays meant the three actors were often sitting around, under-stimulated, waiting for filming to resume. And it was during these long gaps that tensions started to run high. The conflict between Shaw and Dreyfuss, in particular, would later become known as one of Hollywood's greatest ever feuds.*

Dreyfuss described Shaw as 'an enormous personality' in the 2010 documentary Jaws: The Inside Story. *'In private, he was the kindest, gentlest, funniest guy you ever met. Then we'd walk to the set, and on our way to the set he was possessed by some evil troll, who would then make me his victim.'*

Spielberg himself was under an enormous amount of pressure. He brought his own pillow with him from home, and put celery in it, a smell he apparently found comforting. He didn't have time for anything but work, although there were rumours he was conducting an affair at the time, regularly ferrying a friend out from LA. Of course, like many showbiz rumours, this may have been sheer fantasy.

A Runaway American Dream

Dreams weren't meant to be dreamt any more, not in the music business. Dreams were for dreamers, not grafters and career artists. Bruce Springsteen, who was both, sold himself as a dreamer. And if you could wend your way through the hype, you could see why. His third album was a glorious statement of intent: it was *West Side Story* on big steel wheels, a ruthless pursuit of sensation that sounded like Bob Dylan produced by Phil Spector. And no one believed in it as much as the dreamer from Jersey.

Born to Run by Bruce Springsteen

BRUCE SPRINGSTEEN WAS LYING IN his bed at his home in West Long Branch, New Jersey, wrapped in nothing but a cap-sleeve T-shirt and his boxers, when he thought of the phrase 'Born to run'. He liked it immediately. 'It suggested a cinematic drama I thought would work with the music I was hearing in my head,' he said. That music was largely composed of Roy Orbison, Duane Eddy, Elvis Presley, Phil Spector, the Beach Boys and Bob Dylan, the music of his youth that had galvanised young America ten years previously. 'At first I thought it was the name of a movie or something I'd seen on a car spinning around the circuit.'

Springsteen believed like no one else in the power and possibility of rock, as well as its redemptive ability; nearly every song he wrote for his next record touched on the central mythical image of the rock and roll era, the ideas of escape and abandon. 'Born to Run' told the story of young lovers seeking freedom. The narrator invites a girl named Wendy to run away with him to discover what life has to offer.

He craves to escape the town that 'rips the bones from his back' before it's too late. Springsteen's original draft, filled with references to his birth state New Jersey, had over fifty pages of lyrics, verse after verse after verse.

'This was the turning point. It proved to be the key to my songwriting for the rest of the record.' That record turned out to be *Born to Run*, a long form 'studio production' which used the recording studio (largely the Record Plant in New York) as an instrument rather than simply trying to replicate the sound of his live performances. 'I had these enormous ambitions . . . I wanted to make the greatest rock record I'd ever heard. I wanted it to sound enormous, to grab you by your throat and insist that you take that ride, insist that you pay attention – not just to the music, but to life, to being alive.'

Some ambition. Trying to capture the nature of Everyman is a daunting task, but ever since he first appeared, in the early seventies, Springsteen had endeavoured to do it time and time again. By 1975, he seemed obsessed with imbuing the aspirations of small town, blue-collar America with a mystic glow. After two hit-and-miss albums – *Greetings from Asbury Park, N. J.* and *The Wild, the Innocent & the E Street Shuffle* (both 1973) – the album was Springsteen's effort to break into the mainstream; he decided he needed to be more straightforward, using a much-lampooned repertoire of songs extolling the virtues of pink Cadillacs, Jersey girls and gimcrack homesteads. Springsteen wanted to affect the American psyche in a way that Woody Guthrie could only have dreamed of. He had previously been marketed as the new Bob Dylan, whereas his record company now understood he needed to be sold as the next Bruce Springsteen. (This was ironic as when he was first signed to Columbia, few people in the marketing department knew how to spell his name.)

Born to Run, in the words of Robert Hilburn from the *LA Times*, saw him defining 'the struggle in life between disillusionment and dreams' (someone once said that the characters in Springsteen songs have a seemingly bottomless capacity for taking slaps in the face without their faith in the dream being too severely shaken), while Dave

Marsh, of *Rolling Stone*, suggested that Springsteen's approach was 'a refutation of the idea that rock was anarchic rebellion. If anything, his shows were a masterwork of crowd control, an adventure in pure co-operation, a challenge to chaos.' When it was released, the critic Greil Marcus said it was 'a '57 Chevy running on melted down Crystals records.'

'I wanted to craft a record that sounded like the last record on Earth, like the last record you might hear . . . the last one you'd ever NEED to hear,' Springsteen said. 'One glorious noise . . . then the apocalypse. From Elvis came the record's physical thrust; Dylan, of course, threaded through the imagery of not just writing about SOMETHING but writing about EVERYTHING.'

In the grand old tradition of folk singers, Springsteen put himself into the situations and crises befalling his characters, as he himself dragged them into his songs. By upholding the retrograde folk tradition, he was legitimising his purpose: to sing from the streets and place himself in the narrative arc of the great American rebel icon. Automobile imagery was crucial to those early romantic notions of sex and freedom, and the heroes of his mini-parables were poets-come-car mechanics – dreamers and schemers who lingered aimlessly on the low-rise industrial park fringes of society, down in 'Jungleland', 'Thunder Road', 'Spirit in the Night's' Greasy Lake or the infamous 'rattlesnake speedway in the Utah desert'. Most of their time, though, was spent on the Jersey Shore, drinking warm beer beneath rotary fans in dilapidated diners as their fuel-injected suicide machines sat patiently outside, gently humming after a two-hundred-mile journey across a dozen county lines. Jesus, it must have been hard work trying to be this disenfranchised.

The presence of Clarence Clemons, his African American saxophonist, on stage was particularly significant, as he symbolised Springsteen's musical roots, and more importantly his debt to rhythm and blues, giving the E Street Band even more cultural equity. A few hysterical critics would claim Clemons's inclusion in the band was somehow tokenistic, perhaps forgetting that physicality – Clarence was a big man who looked even bigger in his pink suits – and, lest we forget, talent – he was a glorious tenor player – counted for rather a lot when you were in a group. Springsteen famously referenced Clemons when he broke off during his sets to tell one of his stories, hokey little fables that came across as fireside chats, even though he might have been standing in front of 100,000 people. Listening to recordings of them these days, they sound like the kind of thing Jack Black might have summoned up in *School of Rock*, and yet Springsteen delivered them with such sincerity that it's hard to take him to task. In point of fact, it was easy to get the impression that Springsteen's ruling passions were plectrums and sincerity.

At the time, Springsteen claimed he'd never had an image, though it would have been hard to find another rock star so easily objectified, and it was as easy to distil the Springsteen myth from the way he looked as it was from listening to his music. In the early days, Bruce's beard and pimp cap gave him the air of any self-important East Coast troubadour, but by the mid-seventies he was starting to look even more generic: lumberjack shirt, faded blue jeans, leather jacket and motorcycle boots. He perhaps looked his best during this period, with his drawn, gaunt face, V-neck T-shirt and windcheater making him seem more like Jimmy Dean than Bobby Zimmerman.

Nineteen-seventy-five was a big year for Springsteen, especially in the UK. In November, he famously played two dates at the Hammersmith Odeon. The concerts were part of Columbia Records' push to promote Springsteen in the UK following the success of *Born to Run*, which had already generated a ridiculous amount of hype, with the singer appearing on the covers of both *Time* and *Newsweek* the month before. The vast amount of publicity accompanying his appearance in London caused Springsteen to pull down a promotional poster in the venue's lobby proclaiming that 'Finally London is ready for Bruce Springsteen and The E Street Band'. 'My business is SHOW business not TELLING. You show people and let them decide,' he said in his autobiography.

We were ready, too, although when I first heard 'Born to Run' on Capital Radio, I didn't understand why it sounded *quite* so much like Phil Spector.

Springsteen had auditioned for producer Mike Appel in 1971, with Appel telling the singer to return when he had written some more songs. Springsteen returned a year later, and Appel signed him to a production deal and secured an audition for the legendary producer John Hammond at Columbia Records. The scrawny twenty-two-year-old from New Jersey walked into Columbia Records' office in New York City on 2 May 1972 with an acoustic guitar and sang a few songs in front of Hammond.

Hammond was born rich and, while his aristocratic family had no truck with jazz, he discovered it through the family's African American staff. For him it was not just an epiphany, but also a mission. He espoused it as often as he could, somehow imagining that it would act as a panacea for racial tension. Over the course of his career, he either produced or discovered many of the greatest names in the history of twentieth-century American music, including Billie Holiday, Aretha Franklin, Pete Seeger and, of course, Bob Dylan. At his audition with Hammond, Springsteen played at least four songs, including 'Growin' Up,' and 'It's Hard to Be a Saint in the City.' 'You've got to be on Columbia Records,' was the immediate reaction.

Appel was managing Springsteen as well as producing him, but while he believed in his artist, he was feeling the pressure from the record label to produce a hit. Springsteen was feeling this too, which was why such a lot of planning went into *Born to Run*. Work began on the record in May 1974, taking fourteen months to complete (six of those dedicated to the title track alone). Appel even gave an early version of 'Born to Run' to radio disc jockeys at the end of 1974, nearly a year before it was finished. Nevertheless, the song was received positively and received frequent airplay. Within weeks, it became something of an underground hit. People flooded record stores trying to buy copies of the single, which didn't yet exist, and radio stations that hadn't been on Appel's small distribution list bombarded him with requests for the new album, which also didn't exist.

In Philadelphia, demand for the title track was so strong that WFIL, the city's Top 40 AM station, aired it dozens of times a day. In Cleveland – a working-class town – the DJ Kid Leo played the song every Friday night at five minutes to six, to officially launch the weekend. During the recording of the album, the critic Jon Landau was brought in to help produce, after he had written a piece for *The Real Paper* in which he famously said, 'I saw rock and roll's future and its name is Bruce Springsteen,' having seen him perform in Boston. He would soon take over both production and management from Appel, which would lead to the lengthy court battle that put Springsteen temporarily out of business.

'Cult artists don't last on Columbia Records,' wrote Springsteen in his autobiography *Born to Run*. 'We miss this one, contract's up and in all probability, we'll be sent back to the minors deep in the South Jersey pines. I had to make a record that was the embodiment of what I'd been slowly promising I could do. It had to be something epic and extraordinary, something that hadn't quite been heard before.'

He was attempting to make a masterpiece, which, at the time, was something many artists were trying to do. Competition between the likes of Bob Dylan, the Rolling Stones, Pink Floyd and the Who, between David Bowie, Led Zeppelin, the Eagles, Elton John and Joni Mitchell was so intense that it became almost legitimised. Both fans and critics increasingly expected great things from their idols, and critical musical primacy was as important as success. And Springsteen wanted to join the A-team. As 1975 hurtled along, the charts were full of artists operating on full engines. The market expected it. Pick up a copy of the *NME, Melody Maker, Sounds, Street Life, Rolling Stone, Creem* or any of the shorter-lived music titles at the time and the veneration and expectation was huge. This was art for art's sake, a narcissistic dream with corporate backing, with the audience mostly an abstract concept.

As the consumer base got bigger, and acts started playing arenas, even stadiums, artists started to feel even more elevated. Even more distant from the people buying records, tapes and concert tickets. As rock became big business so it needed even bigger stars to occupy the space. It was almost as if there was now a super-league of rock stars, a VIP lounge full of cover stars and cover stars only. If you weren't big enough to warrant being on the cover of *Rolling Stone*, then who were you anyway? It felt as though rock was on an inexorable rise, an increasingly sophisticated journey in which the destination was a kind of cultural perfection, a place where rock stars were cavorting on a daily basis with Nobel-winning authors, storied film directors, cultural theorists and conceptualists and neo-expressionists.

By the mid-seventies, rock stars weren't confrontational, transgressive or alternative, they were 'important'. They had the star quality that had been bestowed on most rock stars since the age of Elvis, they had the cultural kudos that had been projected on the big stars of the late-sixties' underground and they were reflected in the corporate sheen of the seventies' music industry. If an artist delivered a below-par album, this became a massive news story. What had happened? How did they manage to go so awry? We were looking forward to a masterpiece and they deliver this? What? Seriously? Delivering a mediocre album was tantamount to having an ugly child. What were they thinking? The music press was omnipotent; fuelled by advertising and access, it was impossible to successfully launch an act or break a band without them. Meanwhile, the mainstream press still treated popular music as a specialised subject, a curiosity, and only really became interested when they became a phenomenon. Seeing Springsteen on the cover of both *Time* and *Newsweek* was an anomaly.

Columbia Records spent $250,000 marketing *Born to Run* (they started spending after the album debuted at an embarrassing No. 84 on the charts) and this, combined with radio play, sent the album racing up. Suddenly, Springsteen was a star. An international star, to boot. The London shows were spectacular, although the

accusations of hype made critics cautious, worried they were somehow being hoodwinked.

'After the Hammersmith shows, the cry went up: over-hyped, over-
long and over here, as they'd said of Yanks before him,' wrote the
journalist Michael Watts. '"The Singing Hamburger!",' mocked John
Walters, John Peel's radio producer, with just enough truth in the joke
to give it force. Fast food plus fast cars equals Chuck Berry retread.
Not the future at all, but the nostalgic past.

In 1975, few would have called Springsteen an especially libidinous
performer. His fanbase had always been more male-orientated as he
always appealed to boys who desperately wanted to be older, as well
as men who wished they were younger. Like the boys in Albert Camus'
The Outsider (another male adolescent passion), who leave the cinema
with affected gaits after watching cowboy films, songs like 'Thunder
Road' made men in Golf GTIs think they were driving souped-up
Thunderbirds, and made men working in photocopy shops dream of
cruising down Route 66. His music was the music of the American
West. The sky. The open roads. A big country demanded big music,
and that's exactly the sort of music Bruce Springsteen wanted to
make. Epic. Cinematic. Unashamedly unironic. Never an inadequate
like so many British pop stars, Springsteen was a man whose internal
workings were all on the outside.

The thing that made *Born to Run* so extraordinary was the fact it
was synthesised, that it successfully fused all those influences, and
the result was more than the sum of its parts. Culturally, musically, it
felt like the very apex of post-war American culture; it sounded like all
that had come before it had been fed into a blender. Of course, the flip
side of this at the time was the sensation that it was rather cartoony,
and that all that care and attention into making the most bombastic,
the most authentic and the most 'produced' example of American pop
had resulted in something that was so easy to lampoon.

His fame immediately hit presidential levels at home, whereas
in Britain we were slightly embarrassed about allowing ourselves
to enjoy something so unambiguous, so prosaic, so old-fashioned.

Street Life: Britain's high minded, short-lived version of *Rolling Stone*. *(Author's collection)*

Kraftwerk: rock stars were not meant to look like this in 1975. *(Gijsbert Hanekriit/Redferns/Getty Images)*

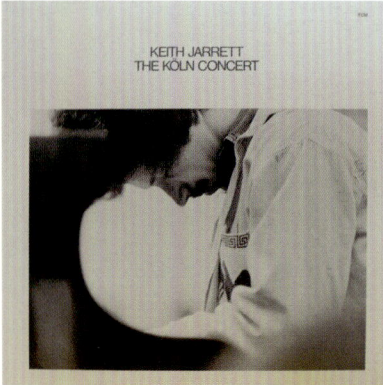

The Köln Concert, which Keith Jarrett inadvertently turned into a global jazz sensation. *(Franz Perc/Alamy Stock Photo)*

Jaws: we're going to need a bigger poster. *(Universal History Archives/Getty Images)*

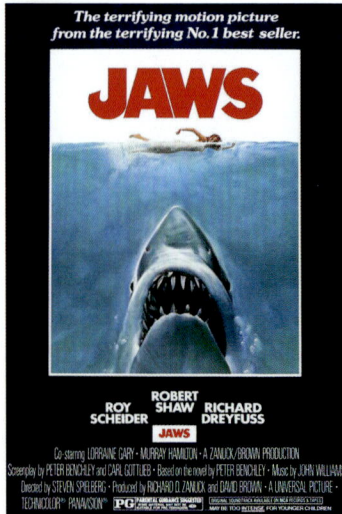

Jack Nicholson was the quintessential seventies film star, whether he liked it or not. And he liked it a lot. *(Floriano Steiner/Ccc/Cipi/Mgm/Kobal/Shutterstock)*

The hub of the Zanzibar was its wavy bar, encouraging convo and high jinx. *(RIBA Collections)*

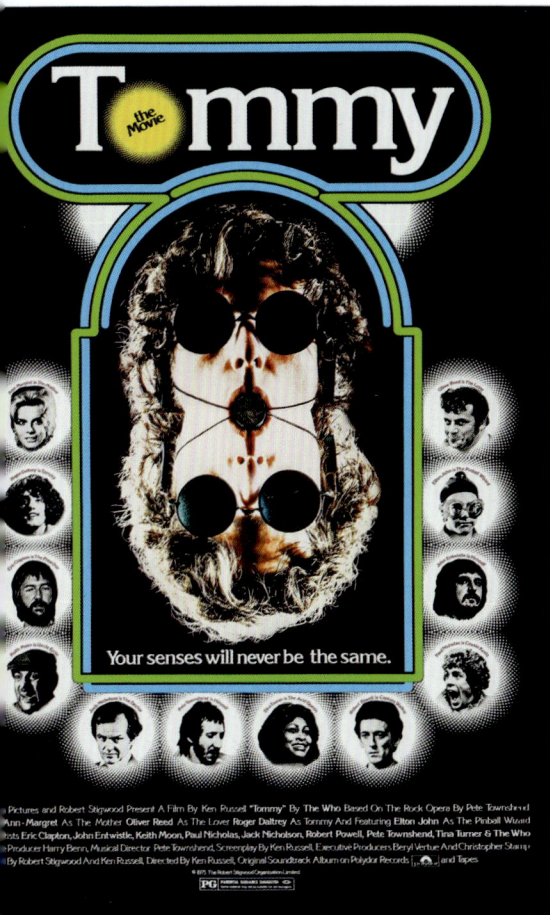

The fall of Saigon temporarily halted the US's post-war imperialism. *(Bettmann/Getty Images)*

Tommy: You'll believe a deaf, dumb and blind kid can fly. *(BFA/Alamy Stock Photo)*

Crosby Street, SoHo, in 1978. Photo by Thomas Struth. *(The Metropolitan Museum of Art/Art Resources/ Scala, Florence)*

Bruce Springsteen in Hollywood: a rare sighting outside New Jersey.
(Terry O'Neill/Iconic Images)

JOWL/JOWL

When David Frost secured Richard Nixon for a mammoth TV interview, the role of the President took on a new transactional hue.
(George G Williams/toonpool.com)

Patti Smith by Robert Mapplethorpe: giving just a hint of what was around the corner.
(Mick Gold/Redferns/Getty Images)

The IRA bombing of the Hilton Hotel made London an international front line. *(Shutterstock)*

Bob Dylan is back! In fact he was backer than Burt Bacharach and his backing band wearing backpacks! *(Vince Maggiora/San Francisco Chronicle/Getty Images)*

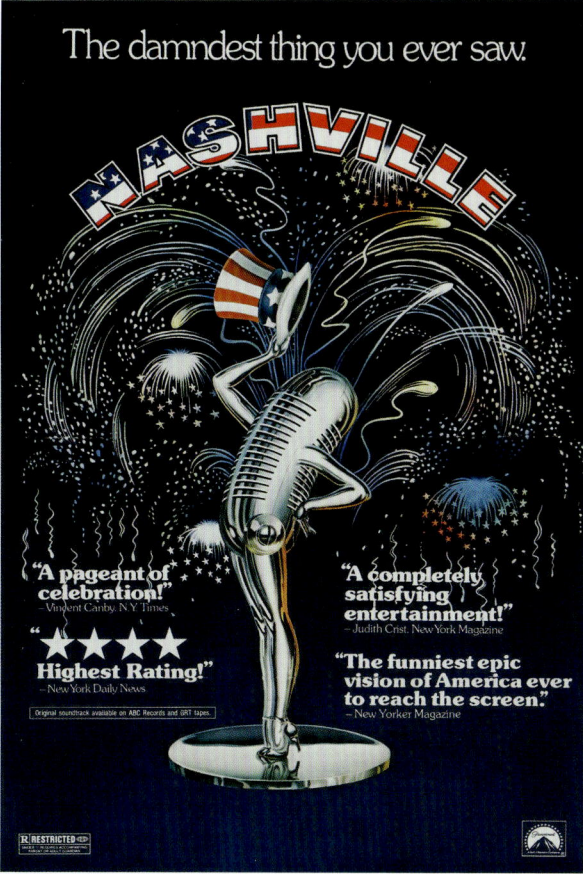

Robert Altman's *Nashville* was an ensemble piece for all of America. *(Paramount/Kobal/Shutterstock)*

You really didn't want to get on the wrong side of Millie Jackson. *(GAB Archives/Redferns/Getty Images)*

The death of Biba and the closure of Nova meant the sixties were finally over. *(Jean-Claude Deutsch/Paris Match/Getty Images)*

There's No Place Like America Today: proving irony isn't always funny. *(Margaret Bourke-White/The LIFE Picture Collection/Shutterstock)*

David Bowie: blue, blue, electric blue. *(Max B Miller/Fotos |International/Getty Images)*

The 1975 Grammy Awards in New York were officially star-studded: Paul Simon with David Bowie, John Lennon, Yoko Ono, Art Garfunkel and Roberta Flack. *(Tom Boxer/Archive Photos/Getty Images)*

If 1975 needed a metaphor, it was *No Man's Land*. (sjtheatre/Alamy Stock Photo)

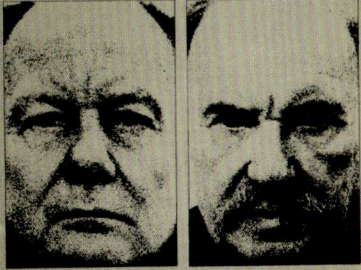

Naked Civil Servants: John Hurt and Quentin Crisp. *Freemantle Media/Shutterstock)*

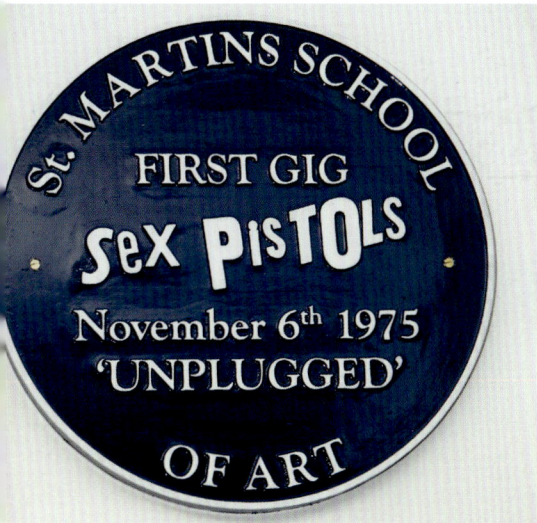

Anarchy in the Charing X Road. *(Dave Etheridge-Barnes/Getty Images)*

Neil Young on Zuma beach. *(Henry Diltz/Corbis via Getty Images)*

Musically, was there a more varied year than 1975? Return to Forever, Donna Summer, Steely Dan, Bob Marley and Smokey Robinson lead the charge. *(Author's collection)*

In 1975, when he was ostensibly still in his larval stage, he used to throw his woolly hat in the air, turn towards the audience, put his hands behind his back and catch it as it fell, just as the band cut the final chord. It was corny but it was more fabulous in the US than it was in the UK. You had to experience America to really experience Bruce Springsteen.

The one thing with Bruce that was universal was his ambiguity. There wasn't any. Bruce was pure, unadulterated, twelve-bar machismo. *Born to Run* is such a distinctive record, imagining a world of impossibly romantic hyperrealism, where the mundane naturally becomes the fantastic. If the depressed state of the Jersey Shore in the early seventies had the fading sense of an era long gone, Springsteen's described an amusement park accompanied by glockenspiel, saxophone and filmic splendour. Springsteen wanted to conjure up the feeling of one, perfect, endless summer night, thirty years of post-war freedom squeezed into one short LP. He wanted the whole album to feel like it could all be taking place in the course of one evening, in all these different locations. He felt the album was built like a tank, so polished that it would be impervious to criticism. Springsteen and his E Street band had spent so much time working on it that by the time it was released he knew that physically, spiritually, intellectually, he couldn't do any better.

'It was indestructible, and that came from an enormous amount of time that we put in, an unhealthy amount of obsessive-compulsiveness,' said Springsteen. 'So part of it was, I was afraid of releasing the record and just saying, "Well, this is who I am," for all the obvious reasons that people are afraid of exposure and putting themselves out there: This is who I am, this is everything I know, this is my best, this is the best I can do right now.'

Before recording a note of either the song or the album, he'd been listening to 'Because They're Young' and a lot of other songs by Duane Eddy because he was into the twangy guitar sound. It was one of those things he couldn't completely trace back. Springsteen had these enormous ambitions. He wanted to make the greatest rock record that he'd ever heard, and he wanted it to sound enormous and to grab you by your throat and *insist* that you take that ride, insist that you pay attention, not just to the music, but to feeling alive, to being alive. That was sort of what the song and album were asking, and it meant taking a step out into the unknown.

'It was a record of *enormous* longing, tremendous longing – that never leaves you,' he said. 'You're dead when that leaves you. It's just about, "Hey, you're gonna take that step into the next day and nobody knows what tomorrow brings." No one can know that. And so the song continues to speak to that part of you – it transcends your age and continues to speak to that part of you that is both exhilarated and frightened about what tomorrow brings. It'll always do that, that's how it was built.'

Springsteen cleverly managed to fuse his own small-town experience with the wide-screen ambition of the record, a sense that this potential life was available to anyone with the guts to try and grab it. And while *Born to Run* was the encapsulation of the rock and roll American Dream – being frustrated, breaking out, and finding short-term pleasure – in essence it had a much stronger, working-class *Grapes of Wrath* sensibility, the American universal. In this he was appealing to everyone, which was a large demographic.

In *Rolling Stone*, he explained the genesis of one of the album's cornerstones, 'Meeting Across the River', which was a prime example of this message. 'There was that New York–New Jersey, big-time/small-time thing. Back then, when you lived in New Jersey, you could've been a million miles from New York City and yet it was always there. By that time, I think we'd been counted out, and it probably had something to do with that, a feeling I had about myself maybe, that you'd been underestimated. Most of the folks that go into my business have had the experience of someone judging your life as being without great value. So that song grew out of, "Hey, that guy's sort of a small-time player, but he's still got his sights set on what's across that river."'

At the time, *Rolling Stone* was an issue for Springsteen, as he felt they didn't take him seriously enough. On 27 October 1975, Springsteen achieved a fairly unprecedented feat of media saturation when he appeared on the covers of both *Time* and *Newsweek*, when those magazines really meant something. Appel had struck a deal with *Time*, where he offered an interview with the singer on the condition he be featured on the cover. They took the bait. Appel extended the same offer to *Rolling Stone*, *Playboy* and *Newsweek*. Only *Newsweek* accepted. The fact that both magazines published their cover stories on the same day added to the hype. Even though the *Newsweek* article was slightly more cynical than *Time*, focusing on the marketing campaign surrounding the release of the album, the twin coverage was remarkable; the last time the same entertainer had appeared on the cover of these two magazines was 28 February 1972, when Liza Minelli was featured for her appearance in *Cabaret*. Springsteen even started referencing it in his shows, altering the lyrics to 'Rosalita (Come Out Tonight)', one of his signature early songs: 'Tell him now it's his last chance, Rosie, Tell him I ain't no freak, 'Cause I got my picture on the cover of *Time* and *Newsweek*.'

But this left a sour taste in his mouth where *Rolling Stone* was concerned, as they had effectively blanked him. He had circumvented them and danced with the mainstream media. 'I was not on the cover of *Rolling Stone* when *Born to Run* came out, you know,' he pointed out later. 'I'm not picking a bone or anything, but I always felt – while we're talking about it – they were a little skittish about putting me on the cover when that record came out.' The magazine's editor, Jann Wenner, countered that the other magazines were 'the establishment' and that Springsteen making the covers of both was a subject of intense controversy.

Springsteen also lamented that such attention attracted an unlikely fan: the Internal Revenue Service. 'I hadn't paid a penny in taxes when I was on the cover of *Time* and *Newsweek*, and the IRS found out about it and it took me ten years [to pay it],' he said.

Of course, many people who came to the London concerts did so purely out of curiosity, intrigued to know whether or not this so-called saviour of rock and roll was anything like as advertised. They came because of the hype, because that's how they were first made aware of him. It didn't matter whether or not he was any good – although the common verdict was a uniformly positive one – he was battling against a backdrop of hyperbole, something that would stay with him for many years to come. In a way this became Springsteen's biggest challenge, with himself, his critics and the public: how to prove he was authentic. How to prove he was more than the sum of his parts. How to prove he was the 'real deal'. This resulted in Springsteen continually overreaching; his career became a constant tussle with perception. Was he the new Dylan? Was he a product of Colombia's marketing department? Was he any better or any more authentic than Bob Seger, Lynyrd Skynyrd or the Allman Brothers? He would turn out to be better than all of them.

In the end, *Born to Run* was everything it was designed to be: a New Jersey rock and roll fantasy that fed on Dylan, Steinbeck and *West Side Story*, an anthemic fun fair of street gangs and broken hearts. In a piece entitled 'Is Springsteen Worth the Hype?' in the *Village Voice*, published on 25 August 1975, Paul Nelson wrote: 'Springsteen's weaknesses stem from too much talent, not too little. When you can achieve just about anything you want onstage it's hard not to stay there until you've rung all the bells; and one often gets the feeling that Bruce is having so much fun he'd gladly pay the crowd to let him do just that. Ironically, if he weren't as good as he is – and he is close to being the best we have – no one would be concerned with such minor issues as pace and overreach.'

It seemed he was worth the hype.

Richard Bernstein's Interview

1 July, 860 Broadway, New York

'*RICHARD BERNSTEIN'S FACES ARE WONDERFUL,*' *Andy Warhol said about the super-stylised portraits Bernstein crafted in his ground floor Chelsea Hotel studio. 'He is my favourite artist because he makes everyone look so famous.'*

By 1975, Bernstein had been drawing the front covers of Warhol's Interview – *the magazine that was the crystal tower of pop – for three years. He had been living in the Chelsea Hotel ever since moving to Manhattan from the Bronx (he would live there until he died, in 2002), and was fully immersed in the giddy social worlds of the city. While he was studying art at the Pratt Institute and Columbia University, Bernstein plunged head first into the budding pop-art movement and its raucous downtown New York City party scene, which fed his appetite for both extravagance and escapism. (He happily took the advice of his professor, who told him to 'work hard, live hard, and fuck hard'.) He partied with the rich and fabulous, sliding precipitously into drugs in the process, then painted his new nightclub buddies when Warhol and his editors decided they were ready for the* Interview *spotlight.*

Interview *had been celebrating bold face names since it launched in 1969, but it wasn't until Bernstein arrived that it developed a discernible style. He turned his cover stars into Mount Rushmore-sized legends, automatically afforded superstar status. 'At the time the magazine looked very intellectual,' said Bob Colacello, who was the editor, 'and it was all black-and-white. In about 1972 . . . we wanted to make it more pop . . . so, Richard Bernstein was brought in.'*

Paloma Picasso, herself one of Bernstein's subjects and daughter of Pablo Picasso, said. 'Richard Bernstein portrays stars. He celebrates their faces, he gives them larger-than-fiction size. He puts wit into the whimsy of fashion and transforms photographs into masterpieces of acceptable kitsch.'

His work was not just emblematic of the high glamour of its time, as celebrity came back into fashion after the countercultural putsch of the sixties, but he was also something of a deep-fake innovator, creating nude images decades before the advent of AI. In 1968, he created one of his most controversial pieces, The Nude Beatles, *a neon-technicolour group portrait with the heads of the Fab Four superimposed on lithe, naked male bodies. The prints were confiscated by order of a French judge and the Beatles label, Apple Records, filed a losing lawsuit against him. Even so, John Lennon was impressed when Bernstein later told him it had been a missed opportunity not to use them on an album cover.*

He produced no-holds-barred illustrations in Newspaper, *a queer art magazine; he famously filled one centrefold with a nude image of transgender Warhol superstar Candy Darling floating naked and supine over plump white clouds.*

For the magazine covers, Interview *would commission a photographer to shoot their cover star before asking Bernstein to turn it into an illustration. Having chosen the image, he would ask for a silver gelatin print which would then be airbrushed into a cover portrait. The cover subject was decided by Bob Colacello, art directed by Marc Balet and photographed by the world's most renowned photographers, such as Greg Gorman, Matthew Rolston, Albert Watson and Peter Strongwater, among others. In all he would paint 189 covers for Warhol, featuring Diana Ross, Clint Eastwood, Lily Tomlin, Mick Jagger, Dolly Parton, Liza Minnelli, Bette Midler, Grace Jones, Marisa Berenson, Jayne Mansfield, Brigitte Bardot etc. He infused the magazine's covers with go-go glamour and created a hyperkinetic, handwritten logo that, as one of his friends pointed out, 'looked as though he was high on coke when he did it' (highly*

likely). It was a time when anyone could be a star, and Bernstein's por-
traits helped catapult them there.

His childhood projects had always been ambitious and flamboyant,
such as a parade float of an enormous goldfish, complete with real
bubbles, built on top of his sister's bicycle. Projects like these were
inspired by Hollywood films like The Wizard of Oz, *which he watched*
obsessively; they fuelled his growing fascination with fantasy and
glamour. Bernstein's parents then enrolled him in Saturday classes at
the Metropolitan Museum of Art for gifted and talented children. It was
there that he was introduced to the art of Mondrian, Picasso and O'Keeffe.

Whether the Interview *cover star was a breakout athlete, an emerging*
actor or a new socialite on the disco scene, Bernstein enhanced their
best qualities, making them look immortal. He would use a variety of
elements to create his cover images, including paint, coloured pencil,
pastel and correction fluid – but it was the puffs of airbrush which gave
his subjects otherworldly auras and the dewy glow of eternal youth. They
simply looked like gods. Each and every one of them. As Grace Jones put
it, 'Richard's art made you look unbelievable . . . You didn't have to go in
already made up for him. He did the makeup on you. And it was magic.'

If the success of Interview *could be construed as an antidote to*
the harsh realities of the sixties – Vietnam, the Manson murders, the
assassinations of Robert F. Kennedy and Martin Luther King Jr. – then
Warhol had been a genius in anticipating the desire for artifice and
glamour. With his magazine's emphasis on bright colours, commodity,
and celebrity, Bernstein's cover images were the perfect advertisement.
His cover stars had charmed lives, much like his readers. This also
chimed with a growing obsession with physical perfection: where
gym memberships, plastic surgery and body worship would become
commonplace. Bernstein's work was the perfect example of Warhol's
maxim that everybody could be world-famous for fifteen minutes.
They certainly could be in 1975.

'He was great at glossing the gloss,' said his friend David Croland, artist
and party regular and onetime boyfriend of both a Warhol Superstar
(International Velvet, née Susan Bottomly) and Robert Mapplethorpe.

'Stars are glossy to begin with. That's why they're stars – they shine a little brighter. And Richard made them shinier.'

The July cover star was Florinda Bolkan, the legend of arthouse and grindhouse Italian cinema.

In the Lands of the Lotus Eaters

If California had reached such an exalted state of gentrified stealth, a global hub now for the music industry as well as the new Hollywood generation, then didn't its denizens deserve their own Boswell? They certainly thought so. And in Joni Mitchell they found one: a diarist whose early songs were a kind of calligraphy of her moods, but who was now writing songs about the rich. The instrumentation she was reaching for in 1975 belied her roots, stretching into jazz, while others were retreating into the West Coast lifestyle.

The Hissing of Summer Lawns by Joni Mitchell

IN *ONCE UPON A TIME World*, his encyclopaedic history of the Côte d'Azur, Jonathan Miles writes about the lemon-scented town of Menton, which liked to claim its citrus trees were truly a gift from Eve. 'Expelled from paradise for eating the forbidden apple, the mother of us all grabbed a lemon and – wandering over the Earth – threw it down in the countryside near Menton,' he writes, 'where it created a new Eden. It wasn't Paradise, but the weather was good. Unlike Eve, many visitors to the region arrived not with lemons but with oodles of their own forbidden fruit.' As they soon would in a new Valhalla: California.

Spanish missionaries first planted orange trees in California in 1769, the first commercial orange orchard was planted in 1841 near what is now downtown Los Angeles and the fruit soon became a symbol of the Golden State's exponential prosperity. But what always follows prosperity is decadence, whether it's on the West Coast or the French Riviera.

Demand for California citrus grew, along with its acreage, as citrus quickly became the economic base for the Golden State. The boom spurred California's second gold rush – only orange was the new gold. Capitalising on the image of California as the land of opportunity and sunshine, the sector became the first to use advertising to promote an agricultural commodity and the orange became the perfect symbol for the sun and the Golden State. Long before the palm, Hollywood, the Beach Boys, or California Girls of any generation, it was the humble orange that really made California shine.

By the early seventies, California was the holy grail of the lifestyle revolution, and the place where changes in human behaviour – how we ate, laughed and thought about our bodies – were first noticed. New York might still have been the place to go to make it, but California was always a more attractive proposition, a land of prosperity, a place of abundance, where the routes to success were more obviously straightforward. It's where you went if you were from London, New York, or indeed Canada. It wasn't just the land of lotus eaters and OJ (and, much later, O. J.), it was the freshly minted frontier of sex, drugs and rock and roll. Los Angeles fostered a musical revolution; San Francisco reinvented the idea of sexual revolution; Gold's Gym and the muscle pen of Venice Beach became the model for health clubs everywhere. Alice Waters – using fresh ingredients from places such as the Corti Brothers supermarket in Sacramento) transformed the American palette with her Berkeley restaurant, Chez Panisse; the San Fernando Valley became the capital of a booming global trade in pornography; Levi Strauss created a denim revolution; Governor Jerry Brown succeeded Ronald Reagan; and Joni Mitchell came here in 1967 after a peripatetic adolescence that had seen her live everywhere from Saskatchewan and Calgary to Toronto and Detroit. In short order, she met David Crosby – riding high in the Byrds, one of the most famous bands in the US – signed to Reprise Records and became a star.

There is a misconception that many people who started becoming famous towards the end of the sixties somehow got there through altruism, because they actually deserved to be there or simply because they were lucky enough to be around at the time. It's easy to forget that the Love Generation and many who came in its wake were desperate to get noticed. Joni Mitchell was certainly driven, spurred on not least by the polio she contracted at the age of nine. Bizarrely, this was the same Canadian epidemic of 1951 in which Neil Young, then aged five, also contracted the virus. The drive for recognition was one of the reasons Roberta Joan Anderson was far happier using the name Joni Mitchell after she married the folk singer Chuck Mitchell in 1965. The same year she gave birth to a baby girl and, being unable to support her, almost immediately gave her up for adoption (and who she would later seek out, and happily reunite with). Joni Mitchell had a mission.

'Ask anyone in America where the craziest people live and they'll tell you California,' she said. 'Ask anyone in California where the craziest people live and they'll say Los Angeles. Ask anyone in Los Angeles where the craziest people live and they'll say Hollywood. Ask anyone in Hollywood where the craziest people live and they'll say Laurel Canyon. Ask anyone in Laurel Canyon where the craziest people live and they'll say Lookout Mountain. So, I bought a house on Lookout Mountain.'

She developed quickly, moving the singer-songwriter genre on in a way never done before. She made sure she knew the right people to make herself heard, wrote a song about the Woodstock festival without actually being there and hitched herself to Graham Nash (which, in the late sixties, was the equivalent of hitching yourself to Mark Zuckerberg). Her talent was so immense it remade the social space around her. Everything Mitchell did between 1971 and 1979 was *sui generis* and, while she made many great albums throughout her career (including a few late in life, when the voice had become genuinely world weary), the only three anyone really needs are *Ladies of the Canyon* (classic happy Caran d'Ache folk-rock, from 1970), *Blue* (plaintive confessional ballads, from 1971) and *The Hissing of Summer Lawns* (where jazz reared its ugly head, in 1975).

Mitchell's early work advertised a career of ambiguity, with your head going in one direction and your heart in another, as she was never as wholesome or as straightforward as someone like Carole King, for instance. She soon became the high priestess of Laurel Canyon, the West Coast folk heroine, a woman whose talent was easily the equal of any man. As the *New Yorker*'s Dan Chiasson said, 'Men often wanted Mitchell to be a wife, a muse, a siren or a star. Instead, they got a genius, and one especially suited to deconstructing their fantasies of her.' Or, according to Elvis Costello, 'It's not just that she has no rivals among female singer-songwriters. She has very few peers among any songwriters.' She made the best music of her generation by falling in love, over and over, while defending her sense of self.

Hissing . . ., meanwhile, was preoccupied with the innate wildness beneath the suburban surface and it was directed pertinently at the haute bourgeoisie. 'Listen to the slave ships making you clothes, look at the pimps and hos who toil while you sleep, look at the Boho Dance, where artists are frauds and critics are sophists,' said the journalist David Yaffe, and he was right. Lyrically, *Hissing . . .* marked Mitchell's divorce from her past and break from 'personal' songwriting, with the confessional abandoned now for the observational. When it appeared without so much as an 'I' in the lyrics sheet, a sceptical *Rolling Stone* review by Stephen Holden suggested Mitchell had entered 'the realm of social philosophy'. There were no secrets on this album, no admissions, no pop private ennui. The lyrics concern the inner world of society women, flighty women, subjugated women. The album has a cast of unfulfilled women playing out their emotions in a world sodden with cocktails and smudged eyeliner and accessorised with suburban dreams, a peek behind the curtain of palm trees that protected the rich and the bored. Female agency is a central theme of the album, which was released in the midst of second-wave feminism. The more adventurous she became, the wider her canvas got. At the time, she ran with the culture. As Yaffe wrote, '*The Hissing of Summer Lawns* was released in the year of Tom Wolfe's "Me Generation". *Hejira* came out in the year [1976] of Gail Sheehy's bestselling *Passages*. *Don Juan's Reckless Daughter* was released in the year of *Annie Hall* [1977]. Joni was telling her listeners who and what they were, and as they were making their way through their self-discovery and discursive relationships, Joni's personal expression was part of the larger conversation.'

Me? I couldn't believe how sophisticated it was. It sounded like a book. Her work was full of artifice, and yet it was purposefully less demonstrative than most of the work by her male peers. That much was obvious.

Mitchell was taking pop to a new level, using the kind of lyrical density usually found in the Great American Novel. 'Its musical vocabulary – as well as its lyrical one – fell magnificently between acoustic realism and symbolic fantasy,' wrote Winston Cook-Wilson in *Pitchfork*. 'The album's final track provides an apt conclusion in every sense. It's a bare summation of the threads of struggle, compromise and perseverance that run through the record, but also eerily evocative of its fate in the world.'

Musically, the record was beyond mature; if *Court and Spark*, her previous record, had been steeple-fingered luxury pop with an awareness of jazz (and where she began her transformation from memoirist to a writer of imaginative short fiction), *Hissing...* was a whole new world. This was a harmonically rich, musically complex and intellectually formed piece of work that not only gave a vivid picture of where Mitchell's ambitions lay, but also showed the extent to which 'pop' or 'rock' music had changed since the end of the starburst sixties. This record sounded very much like a grown-up work made by grown-up people for other grown-ups to listen to. In 'The Boho Dance' she referenced Tom Wolfe's *The Painted Word*, his pointed critique of the abstract expressionists, minimalism and the avant-garde. The cover was her own Rousseau-like painting and she alluded to literature everywhere. Indeed, it was a prime example of early AOR.

The quality of the music on display is formidable. Mitchell employed the band that had accompanied her around Europe in 1975 – Tom Scott and the fabulous L. A. Express – augmented on various tracks by two other lead guitar players, Larry Carlton and Skunk Baxter. She roped in David Crosby, Graham Nash and James Taylor on backing vocals, as well as the East African Burundi drummers. The record was idiosyncratic, but it was slick.

The sophistication of the music on *The Hissing of Summer Lawns* more than matched the adult nature of her lyrics. In fact, at the time, you couldn't find a more adult writer in the rock idiom. Certainly not among her male peers, who were intent on consolidating themselves as sexy seers of the ages. Here was an often searing example of one of the prime motivations of mid-seventies' rock: to make music for adults, music that grew old with its audience. *Hissing* . . . was so sophisticated, it wouldn't have been surprising to be handed a gift with purchase as you left the store clutching it – maybe a little plastic cup containing a selection of frozen grapes for the journey, as you climbed into your Mercedes in the Tower Records parking lot.

In the late fifties and early sixties pop had been all about being a teenager, centred around songs concerned with teenage sensibilities and teenage concerns. And once the late-sixties' freak flag had been unfurled, the music industry in the early seventies returned to the same territory. Rock, however, was about something else completely, especially Joni Mitchell's kind of 'rock'. Not for her the rich complacency of the Eagles or Crosby, Stills & Nash. Sure, there were palm trees in Joni Mitchell's garden but they cast long shadows. Here on *Hissing* . . . were the complexities of the evolving adult world, accompanied by saxophone and fretless bass. Her audience thought they were sophisticates, and they demanded the kind of music that reflected that and flattered them. And here it was, all manicured, smug, and heat-sealed. It's been said that every song on the album illustrates the feeling of being trapped, unworthy and marching to the obscure beat of someone else's drum, while 'Edith and the King Pin' is easily as adroit as a John Cheever short story or a chapter in an Anne Tyler novel.

But the sound, the sound . . . Mitchell's album was one of those records, like the Eagles' *One of These Nights*, released in the summer, five months earlier, that acted as a soundtrack to California, conjuring up images of tall, swaying palms, wide hot tarmac, and sunglasses and bikinis, and the perpetual widescreen mythology of hot love and divine success. At the time, there were few greater joys than being in an open-top car and stumbling on the perfect radio station, the one that played the best driving records, as you drove back to LA, speeding though the desert with the wind blowing through your hair. FM rock music at the time tended not to be ironic, it wasn't post-anything, didn't have to be self-referential or cute, didn't need to reference the past . . . it was just rock music, the sort that just sounded like it was escaping its past, not responding to it. This was iconic stuff, the stuff that conjured up visions of dark, desert highways, gargantuan palm trees, Cinemascope skies and blithe hitch-hiking honeys in cut-off jeans and too-tight T-shirts. The Big American Car Music of the time – the Eagles, Doobie Brothers, Steely Dan, Lynryd Skynyrd – sounded like it was built for the journey (i.e. built for comfort rather than speed). American cars never had gears, and rarely did the best Big American Car Music – it was either all in fifth, or it automatically changed without you noticing, moving effortlessly from verse to bridge to chorus to coda as the highway cut a swathe through the landscape. Joni Mitchell was a part of this, but if she occupied a position in the radio schedule it was always at night, during the contemplative hours when the audience wanted ballads and retrospection rather than urgency and hope. And Joni did this better than anyone.

By the end of 1975, Joni Mitchell's seventh album had interfered with the cultural hegemony of Los Angeles. Look quickly and you'd think she had conquered the West Coast media yet again; look more closely and you saw she was starting to be accused of overreach. Stand back and give it some macro and you'd see quite quickly that her principal contribution to the culture in 1975 was elevating the role of the rock album in the development of twentieth-century culture. Because rock stars were being indulged to such an extent that every proclamation was treated with an undue amount of respect by an indulgent public, they began to agree with them. Some of her audience found the album difficult, as did critics. Honestly, it was like Dylan going electric all over again. The record was easily as good as any that came before it and, in reality, was actually a good deal better, but the jazz optics irritated people. They liked the boundaries of her earlier work and questioned why Mitchell had to move forward quite so quickly. Ironically, for music that on the surface sounded so safe, she was developing in a way that would start to unnerve her audience. America liked copycats; she tried to make music that was enduring and was a recipient of prejudice as a result. Was it because she was a woman?

In 1979, in an interview with Cameron Crowe, Mitchell questioned why Steely Dan could get away with working so closely with jazz musicians on *Aja* (which came out in 1977), whereas she was lambasted instead of being applauded. *Hissing . . .* was an impeccably produced album that would turn out to be her jazz-rock zenith, everything delivered with dazzling thrift, like a collection of perfectly dramatised short stories. 'Mitchell's people vacation in Paris, shop at Bloomingdale's when they are in New York, party at night in the slickest discos,' wrote the reviewer in the *Detroit News*. 'Some of them can even claim to have "a room full of Chippendale that nobody sits in". Unfortunately, in delineating the ennui and emptiness that lie behind this brave material front – and if that idea seems familiar, if you think you've read it in a book before or seen it in a film somewhere, you're right again – Mitchell takes a tone that is smug, sometimes so smug that it is downright irritating.'

'*The Hissing of Summer Lawns* is a suburban album,' said Mitchell. 'About the time that album came around I thought, I'm not going to be your sin eater any longer. So, I began to write social description as opposed to personal confession. I met with a tremendous amount of resentment. People thought suddenly that I was secure in my success, that I was being a snot and was attacking *them*. The basic theme of the album, which everybody thought was so abstract, was just any summer day in any neighbourhood when people turn their sprinklers on all up and down the block. It's just that *hiss* of suburbia.'

In this she was being slightly disingenuous, as Mitchell's idea of suburbia was different from most people's: gated communities full of big cars, kidney-shaped swimming pools and Latino gardeners alongside the relentless sound of sprinklers. The song itself was inspired by a visit to José Feliciano's house, with Mitchell being struck by the manicured, irrigated, bottle-green lawn outside. She reimagined it as a virtual prison for someone's wife. In Mitchell's imagination, her lawn became a symbol for a relationship in which luxurious trappings are precisely that: designed to keep a woman both pliant and captive.

Hmmm. She thought it was odd that people called her narcissistic for swimming around in a pool, which she was photographed doing on the album's inner sleeve. She said it was an act of *activity* as opposed to sexual posturing and was angered by the fact her fanbase wanted 'more morose, scathing introspection'.

She had a point. It wasn't just the jazz or the supposed narcissism, ironically it was the way in which she had turned her sights on other people rather than herself. In literature, in the world of books, the exploration of the pampered, educated, but troubled middle-class male, by the likes of John Updike, Philip Roth, Joseph Heller and John Cheever, was a well-ploughed furrow. They were men who were exploring their inner conflicts in the way singer-songwriters had started to do, using their particular experiences as universal truths. As soon as Mitchell started writing her own universal truths, the knives came out. Even John Lennon was bitchy: 'I played him something,' Mitchell later recalled. '[He said] "Oh, it's all a product of overeducation."'

'Here's the thing,' she said. 'You have two options. You can stay the same and protect the formula that gave you your initial success. They're going to crucify you for staying the same. If you change, they're going to crucify you for changing. But staying the same is boring. And change is interesting. So, of the two options, I'd rather be crucified for changing.'

The rot would set in soon enough. Two years after *Hissing . . .*, in a move that would be widely ridiculed and condemned, Mitchell pushed the boundaries of good taste by appearing in blackface on the cover of her sprawling 1977 album *Don Juan's Reckless Daughter*. Complete with an Afro wig and a joke moustache, Mitchell took on the persona of 'Art Nouveau', a hipster character who combined some of the more stereotypical elements of seventies' black culture, including a pimp suit and gold jewellery. Mitchell originated the character during a Halloween party in 1976. 'There were a lot of people out on the street wearing wigs and paint and masks, and I was thinking, What can I do for a costume?' she said. Days later she was tending to her teeth. 'I was being butchered by a dentist who was capping my teeth – and he was my dentist for about twelve years and one day he said, "Oh, you've got the worst bite I've ever seen. You have teeth like a Negro male,"' she said. Mitchell then said that after leaving the dentist, she saw a black man strutting in what she called 'diddy-bopping', before he allegedly turned to her and said, 'Mmm, mmm, mmm, you looking good, sister, you looking good.'

'His spirit was infectious, and I thought, I'll go as him,' said Mitchell. 'It was as if this spirit went into me. So I started walking like him . . . I bought some pancake make-up. It was like "I'm goin' as him."' She said she felt she shared an affinity with black men. 'When I see black men sitting, I have a tendency to go – like I nod like I'm a brother. I really feel an affinity because I have experienced being a black guy on several occasions.'

As self-absorption and age started to infect her spirit, the zeitgeist moved on, leaving Joni Mitchell's still-beating heart of the decade's West Coast cultural renaissance lying on the sidewalk at the cross-section of Hollywood and Vine. But that lay in the future. In 1975, *Hissing* . . . was the consensus masterwork of an eclectically brilliant composer and assured musician and singer, a simply beautiful piece of work. *The Hissing of Summer Lawns* remains the high-water mark of seventies' rock music, an exotic and challenging piece of work that immediately announced itself as 'important'. This was an emotionally complex work that demanded to be taken seriously, with songs of such sophistication, such intricate lyrics and beautiful melodies, they were bewildering. Juxtaposing the primitive and the sophisticated, both on the cover and on the record itself, she played with the spiritual bankruptcy of America's upper class while inadvertently exploring the tension in her appropriation of jazz.

Her songs were vignettes: for instance, 'Edith and the Kingpin', which must rate as one of her very best songs, a compositional tour de force ('Part of it is from a Vancouver pimp I met,' she said, 'and part of it is Edith Piaf. It's a hybrid, but all together it makes a whole truth'). As Mitchell said herself, with this album she moved away from the hit department to the art department, creating a strange, rather cynical masterpiece in the process, a dark, yacht rock record that was both progressive and transitional. A bridge between her past and her future, between folk and jazz, between the confessional and observational, between the mysterious and the ambitious, and brought to life with extraordinary precision and flair, *The Hissing of Summer Lawns* was the personification of haute bourgeois pop. In short, the album was a major work of art. Which was *very* 1975.

Cowboy Country

8 August, Martin Theatre, Nashville

BY THE TIME HE GOT around to making Nashville, *Robert Altman had already proved himself to be something of a maverick of seventies cinema by delivering* M*A*S*H, McCabe & Mrs. Miller, The Long Goodbye *and* Thieves Like Us. *While* Nashville *would be his apotheosis, each of his films had been infused with a particular seventies' sensibility of rambling exploration, of a decade that felt as though its boundaries were already fading. Along with Martin Scorsese, Francis Ford Coppola, Paul Schrader, Steven Spielberg, Brian DePalma and the rest, Altman was using movies as a way of continually defining the decade – sometimes in anticipation, sometimes sweeping up behind.*

Altman had already prepared us for actors who didn't really look as though they were acting, had already attuned us to the comic subtleties of a multitrack sound system with overlapping dialogue that made the sound more 'live' than it ever had been before; he'd even evolved an organic style of movie-making that managed to tell a story without using a traditional narrative. But Nashville *was the quintessential Altman experience, a disaffected portrait of bicentennial America.*

Nashville *was a multi-textured satire following twenty-four main characters involved in the country and gospel industry in Nashville, Tennessee, over a five-day period leading up to a gala concert for a populist outsider running for president on the 'Replacement Party' ticket. It starred Ned Beatty, Karen Black, Keith Carradine, Shelley Duvall, Geraldine Chaplin, Lily Tomlin, Scott Glenn, Jeff Goldblum and dozens more. The impressionistic and seemingly random nature of the narrative successfully evoked the chaos of life in the city.*

The film had a highly complex, subtle structure which was carefully created to produce maximum thematic resonance. The main characters' lives intertwine over five days of political rallies, recording sessions and personal dramas in a rich tapestry of Americana that captures the zeitgeist of the period with unflinching honesty and wit. It was dense, funny, and some said it was a masterpiece. Many of the actors ad-libbed their lines and even wrote and recorded their own country songs. Altman didn't think the film was especially commercial, even though it was meant to be. 'A film like Nashville *is not about the words that are said,' he said. 'The dialogue that is used is part of the behaviour of the actors. It is very important, but I don't think it makes much difference whether one says this line or that line.'*

The film was, in short order, a Grand Hotel-*style extravaganza; a country and western musical; a documentary essay on Nashville and American life; a meditation on the relationship between the audience and the performer and an Altman free-for-all.*

The director is usually remembered as part of the New Hollywood wave of the seventies (that radical flowering of the American film establishment in the late sixties and early seventies); he was actually part of the generation of Stanley Kubrick, Sidney Lumet and John Cassavetes, chronologically at least. He was forty-four and had already made five feature films before M*A*S*H *bought him real fame. But because of his style and his sensibility – the drifting camerawork, the decentred shot composition, his verité recording techniques – he became so closely associated with the decade that in future he would always look out of step. He said he was an instinctual filmmaker, trying to make sense of the vague, smokey things in his head that somehow informed what he was trying to do in front of the camera. 'You took all these elements and then put them together again. I guess that's the art of editing,' he said.*

'If he disdained narrative, he loved situations, using plot more as a way of throwing his characters together in various combinations

rather than a unifying thread,' according to the Harvard Film Archive. 'With his combined love of and derision for tradition, his assertions of individuality together with his need for community, his mixture of high and low, and his alternations between delicacy and crassness, Altman seems a uniquely American figure.'

When you went to see his films, it was less like going to the cinema and more like going to live somewhere else for a while. And with Altman you could never anticipate where that might be. After he had been vindicated by M*A*S*H, *rather than use it as a springboard to commercial glory, he routinely chose unlikely projects, bucked studio authority and developed a working method that delighted some collaborators (particularly actors) and alarmed others (including many writers, some executives and sometimes even audiences).* Nashville *was an absolute delight and after watching it (the movie was nearly three hours long) you immediately wanted him to cast his eye over similar scenarios in Los Angeles, New York, Berlin and London, a never-ending contemporary grand tour.*

His profoundly imaginative disregard for Hollywood conventions and genres meant that any cinematic orthodoxies were ignored, replaced by such a sense of freedom that the audience was propelled forward, even if the storytelling was oblique. Altman also had an affectionate understanding of loners and outcasts, perhaps as he felt one himself. Squint, and you could call him the perfect seventies director, as his documentary style fitted the way in which many parts of America were developing; after the fantastical changes of the sixties, this new decade almost felt entrepreneurial in comparison, with a generation trying to build on the freedoms created by the previous decade.

Essentially, Altman had a rebel spirit, and was so particular that he even spawned his own adjective, 'Altmanesque'. At the time, Pauline Kael was the film critic of the New Yorker, *and what she said about a movie helped enormously with those pictures that weren't going to be on everyone's radar. She loved* Nashville *and write a eulogy to it as soon as she could. 'You don't get drunk on images, you're not overpowered – you get elated. I've never before seen a movie I loved in quite this way: I sat there smiling at the screen, in complete happiness. It's a pure emotional high, and you don't come down when the picture is over; you take it with you.' Everything she said was true, and if there was one film that encapsulated 1975, it was this one.*

Eins, Zwei, Drei, Vier!

During the depths of the cold war, West Germany's economic miracle was set against the social radicalism of the post-1968 counter-culture and the avant-garde of Beuys and Stockhausen. The unifying factor used tech as a way to interpret the future. Krautrock groups explored ways in which their music could dovetail with the wider cultural narrative. In their attempt to move away from Anglo-American R&B roots, melody was subsumed by structure, structure by noise.

'75 by NEU!

GERMAN MUSIC WASN'T IN THE rudest health in the early seventies. On one hand there was a plethora of derivative third-division rock bands, and on the other the oompah rhythms, accordion accompaniment, and nursery rhyme lyrics of schlager.

Schlager – once described by the *New Yorker*'s John Seabrook as what American pop might have sounded like without any African influence ('cheesy', in a word) – was already something of a national embarrassment, the kind of boisterous, child-like pop that consisted of little but saccharine melodies and inane lyrics. It was popular in Weimar Germany and returned triumphantly after the Second World War in German-speaking nations and Scandinavia. The genre's most successful exponents were probably ABBA, or at least the ABBA that once sang 'Waterloo', 'Ring Ring' and 'Honey Honey'. German pop was the kind that competed, without irony, in the Eurovision Song Contest. 'It was considered arch-conservative and bourgeois,' the ethnomusicologist Julio Mendívil said. 'Schlager evoked consoling images of a beautiful and romantic countryside and painted a distorted, idyllic picture of Germany that excluded any reference to the horrors of the war.' In 1975, schlager had yet to be afforded the re-evaluation of irony and was therefore still fantastically annoying. German pop sounded backward, sometimes tainted by the country's history.

German rock music was, if anything, even worse, as it felt imported. The list of German bands trying to break into the mainstream at the start of the seventies was not exactly crowded with talent. The accepted method as the time was to copy British and American groups as diligently as possible, in terms of both presentation and content. Many of these groups were simply covers bands, while many of them looked and sounded a lot like the double-denimed Free. There were, however, a lot of them: Birth Control, Lucifer's Friend, Epsilon, Gift, Grobschnitt, Hardcake Special, Lilac Angels, Metropolis, Panther, Parcival, Thirsty Moon, Streetmark, Triumvirat, Anyone's Daughter, Octopus, Bastard, Pancake and Straight Shooter, to name only a few.

It was in this environment that krautrock began to develop, conjuring a new sound out of the remnants of the pockets of cultural insurrection of the sixties. The genre started to be appreciated in Britain, too, by a handful of journalists, musicians and DJs.

At the beginning of the seventies, it wasn't always easy to listen to the kind of music that was being written about in the music press. If you read about a new band in *Sounds*, *Melody Maker* or the *New Musical Express*, your only chance of hearing them on Radio 1 was after 10 p.m., on programmes hosted by Bob Harris, John Peel and Annie Nightingale. Peel's almost pathological curiosity meant he was pre-programmed to seek out the difficult and the esoteric, and in the process educated an entire generation in everything from pub rock and reggae to prog and folk to electronica and krautrock. He was a huge fan of the German duo NEU! (pronounced '*noy*', German for 'new!'), which in the hallowed halls of the BBC's Broadcasting House would have made him unique.

NEU! were guitarist Michael Rother and drummer Klaus Dinger, whose sound was proscribed by Dinger's pulsing, metronomic drumming. Dinger referred to it as 'a feeling, like a picture, like driving down a long road or lane. It is essentially about life, how you have to keep moving, to go on and stay in motion. To be driven by the drive.' Their records did indeed sound as though they were written for a road movie. Only this movie would have German, not American, roads and instead of saturated colour images of Route 66 or Monument Valley, the NEU! road trip would almost certainly have been in black-and-white. The images you saw through the windscreen of the BMW would have exemplified the growth of a new industrial superpower as the ghosts of the country's past lay dying on the verges. There was an insularity about the NEU! noise, and you could almost sense the fog coming in when you put the needle to the record.

During the depths of the cold war, West Germany's economic miracle was set against the social radicalism of the post-1968 counterculture and the avant-garde of Joseph Beuys and Karlheinz Stockhausen. The unifying factor was technology, as well as the idea of using tech as a way to interpret the future. The bands classified as 'krautrock' (a name invented by Virgin Records' Simon Draper as a ruse to sell Tangerine Dream records) or '*kosmische*' ('cosmic bands') – Can, Faust, NEU!, Kraftwerk, Amon Düül II and Tangerine Dream – were all exploring the ways in which their music could dovetail with the wider cultural narrative. In their attempt to move away from the R&B roots of Anglo-American rock music they created a sound fortified by extended improvisation, musique concrete techniques, and early synthesisers. Melody was subsumed by structure, and structure by noise. They represented the idealism and terror of the times, the deep heat of the cold war at the heart of a divided Europe. In NEU!'s case, Dinger and Rother distilled the core elements of garage rock down to a pounding heart pulse, creating a hypnotic sound that echoed the industrial noir of the Stooges or the Velvet Underground. As their music consisted of little but pulse and texture, so NEU! seemed even more reductive, with such insistent, tribal drumming and impulsive, urgent guitar that they were genuinely provocative. As one journalist put it, 'NEU! set the template for what has since become known as krautrock: a hypnotic beat, a simple melody and waves of atmosphere over mostly instrumental numbers that are like a soundtrack for a bright urban future.'

For such a firebrand of German music it's odd to think that Michael Rother grew up in Wilmslow. His father worked for British European Airlines (BEA), and so the family had lived in Munich, Düsseldorf and Pakistan as well as Manchester, with Rother once attending a British military school in Karachi. His father was meant to be taking over the North American division, until he contracted cancer and passed away. 'So that story ended, we stayed in Düsseldorf, and I started playing guitar,' said Rother. The guitar in question was a Japanese banjo, a shamisen, while his influences at the time were the repetitive street music of Karachi, his mother's Chopin recitals and Little Richard. As he got older, he formed a covers band (*'sklavische Kopie'*, literally 'slave copy'), obsessed with British guitar heroes like Eric Clapton, Peter Green and Jeff Beck. Unlike the members of Kraftwerk (who he would briefly join), he could see the possibilities in a German version of Western pop, but as he became radicalised so he moved on, operating a 'year zero' policy and saying goodbye to blues and rock and all the accepted ideas of what a traditional British or American rock band should be. He envisioned a world free from the conventions of pop songwriting, and perhaps music that might be informed by the cold war, the military complex and the never-ending conflict in Vietnam. When he started NEU! he was working in a mental health facility in Düsseldorf.

Dinger studied architecture for three years but abandoned his studies to concentrate on a career as a professional musician. 'I borrowed money to buy myself a new drum kit and I withdrew for six months in order to practise and become more proficient,' he said. 'After that I joined a band called the Smash. We were a cover band really, but we played live a lot, mostly in southern Germany, so at least I could make a living from playing music. Then in the autumn of 1970, I got a call from Ralf Hütter.'

In 1970, Dinger was asked by the Kraftwerk member and his colleague Florian Schneider to work on the band's debut album and, when Hütter temporarily withdrew in 1971, the remaining pair brought in Rother on guitar. Dinger and Rother's experience in this version of Kraftwerk (lasting just six months) proved crucial as the trio's improvised live performances laid the sonic foundations on which NEU! would subsequently build.

At the time, Kraftwerk were unable to successfully reproduce their live sound in the studio, which compounded the aesthetic and personal differences which were starting to destroy the group. Consequently, in late 1971, Dinger and Rother struck out on their own to form NEU!, with the aid of Kraftwerk producer Conny Plank. Their self- titled debut album was recorded over a four-day period that December at Plank's Hamburg studio. To keep costs down, they worked only at night, when the rates were cheaper. Although Dinger and Rother set out with causal, rough ideas, for the most part their creative process embraced spontaneity.

They preferred 'motorik' drumming, the four-to-the-floor beat that had quickly become the bedrock of krautrock. It was basically a 4/4-beat, the most standard of all rhythms in rock, but Dinger gave it an entirely new dimension through his technique. Instead of accessorising the beat, as any other self-respecting drummer would have done, by adding fills and cymbal crashes or creating tempo changes, he quite simply kept playing it, often creating pieces that would last ten to fifteen minutes. In Dinger's hands, it was no longer a beat, but more like a pulse, a very human pulse. One critic said it 'sounded as if it had gone on since the beginning of time and would continue for all eternity. It was like the rhythm of life itself.'

The originators of post-war German music were always at pains to say how they had striven to find something genuinely alternative to Western pop, and yet here was its very own backbeat. 'Motorik' is German for 'motor skill', and it was pioneered by Jaki Liebezeit, the drummer with Can, the style borrowed, perhaps, from the Velvet Underground's Moe Tucker. NEU!'s Klaus Dinger tended to call it the

'Apache beat'. Dinger emphasised that it was 'very much a human beat', adding, 'It's essentially about life, how you have to keep moving, get on and stay in motion.' Dinger would thrash his drums so hard that he would occasionally cut himself on stage, sometimes with a broken cymbal or splintered drumstick, meaning his drums would be covered in blood. He was like a proto-punk Ginger Baker (he admitted to taking more than a thousand acid trips). The often-bizarre combination of Dinger and Rother meant that the group's debut contained some of the period's most inspired and inventive music.

Brian Eno said, 'There were three great beats in the seventies: Fela Kuti's Afrobeat, James Brown's funk and Klaus Dinger's NEU!-beat.' The band's name had a pop-art background, being the most common slogan in German advertising, and after a while it became almost invisible. They were certainly influential.

Their second album came out in 1972, although its recording almost didn't happen. Halfway through the sessions the pair ran out of money and filled the second side of *NEU! 2* with scratch versions of their only single, 'Super/Neuschnee', played at different speeds on a cassette player that 'howled and chewed tape' and on a hand-driven turntable with a jumpy needle. Brazenly, Dinger called the tactic a 'pop-art solution to a pop problem', and it certainly didn't affect their critical standing. 'I was very well informed about Warhol, pop art, contemporary art,' he said. 'I had always been very visual in my thinking.' Money was a constant issue for the pair of them; they were once so unpopular in Germany that they played to an audience of three people, having driven three-hundred kilometres to get to the venue.

In 1975, after a three-year break, Dinger and Rother temporarily buried their differences and reunited for another stab at the sound they had developed with their debut in 1971. Since then, Rother had worked with the band Cluster in the krautrock supergroup Harmonia and Kraftwerk had become internationally famous. Elsewhere, Can and Faust were redefining the genre and both David Bowie and Eno had started to explore synthetic futurism. The world was getting wise to their sound. By now, Rother and Dinger had diverged in their musical intentions, Rother preferring the ambient and Dinger preferring a more aggressive, rock-influenced style. As a result, they agreed to a compromise: side one was recorded in the old NEU! style, as a duo, with Dinger playing drums. For the pieces on side two, Dinger switched to guitar and lead vocals, recruiting his brother Thomas and Hans Lampe to play drums (simultaneously) – one side an ocean, the other an engine.

The result was essentially a split record with a split personality. The *NME* critic Andy Gill said NEU took 'the repetitive pulse of rock, stretched it out and wove lush, shifting layers of sound over the framework to produce a distinctive, hypnotic music which, whilst undeniably rock, was definitely outside-looking-in . . . This strain reaches full fruition on *NEU! '75*, and it's because of this that the album's so pivotally important.'

Rother said he always felt that Dinger and he were like two action painters in front of a canvas, reacting spontaneously to the contributions of the other. 'The problems Klaus and I have with one another cannot be separated from our music,' said Rother. 'We have such completely different personalities. The actual mystery is how we were able to do the three albums together at all. Our opposing characters sometimes led to great friction, crazy struggles and contradictions in our music. This is what made NEU! so special.' They both believed in repetition, though, which had always been at the heart of NEU! One of Rother's first memories was being a small boy nestled in his mother's arms, engrossed in a story that never appeared to end, that just kept going back to the beginning, in a circle. Rother was fascinated by the story because he didn't know where it started or where it finished.

NEU '75 was very much the sound of the future. This was such a modern record and would be one of the prime reference points for David Bowie's next three albums, *Station to Station, Low* and *'Heroes'*. 'I was completely seduced by the setting of the aggressive guitar-drone against the almost-but-not-quite robotic/machine drumming,' said Bowie. 'Although fairly tenuous, you can hear a little of their influence on the track "Station to Station".' In 1976 Bowie would telephone Michael Rother to tell him he and Eno had been listening to his records, were disappearing to Berlin to start a project, and asked him to join them. Having had an unsatisfactory call from Bowie's manager regarding the contract, and – saliently – money, Rother declined. The project was *Low*.

If Kraftwerk would go on to define and refine the modern parameters of electronic music, both in terms of sonics and optics, NEU! would have a broader and more insidious influence. These may have not been the kind of records that grabbed the public's imagination, but they soon became part of the post-punk playbook. While Bowie and Eno were already starting to work out just how Rother and Dinger had made their records, they would soon become the blueprint for so much of the electronic music that emerged in the wake of punk: it's impossible to imagine the recording careers of Wire, PiL, Joy Division, Iggy Pop, Pere Ubu, Devo, New Order, Einstürzende Neubauten, Sonic Youth (who named a song after them), My Bloody Valentine, Radiohead, U2, Primal Scream, Hot Chip, Paul Weller, Red Hot Chili Peppers, LCD Soundsystem, the Killers, Tall Firs and Stereolab (who can sometimes sound like a NEU! tribute band), without the NEU! prototype. It wasn't just their music, but their sensibility, too, or at least the one imagined by an ideology-obsessed post-punk fraternity.

Dinger had espoused a kind of situationist sensibility, enjoying a neo-dadaist spirit that wallowed in the random nature of the moment. Along with the ambient sounds, the motorik drumming, synthesiser and razor-edge guitar, the pair incorporated everything from electric drills to household objects. This was bricolage by another name and they made sounds, not stories. This stretched to their graphics, too, which predated punk (and the likes of the Sex Pistols' Jamie Reid) by at least five years. Dinger's cut-and-paste montages looked like they'd been put together in a matter of minutes with the aid of a typewriter, some sticky-tape, some felt-tips and a photocopier.

In three years' time, if you had walked into Rough Trade in Notting Hill, the shop would have been full of records that both looked and sounded like NEU! Rother would go on to form Harmonia – who Eno described as 'the world's most important rock group' – and would become one of the most influential musicians of the decade, a guitar-wielding polymath whose work epitomised the revolutionary spirit of the German avant-garde. Not that anyone would have suspected this in 1975. Not even Rother himself. 'It's much better than the situation I experienced in the seventies,' he said in advance of a concert at the Barbican, in London, in 2024. 'I was totally convinced of what we were doing, but people just ignored us or hated the music. So I try to stay independent of praise and criticism. Deep inside, I think I'm not as good as some people think. But I'm not as bad as the others think. So I keep my own opinion.'

Dinger died in 2008, three days short of his sixty-second birthday. 'He didn't put on brakes,' said Miki Yui, his widow. 'He wasn't moderate. He didn't care about health, he did what he wanted, he was a real rock star kind.'

Paul Weller, a man who still wasn't famous by the time of NEU!'s third record, says, 'Michael and Klaus are true originators. Fifty years ago they were making the sound of tomorrow.'

David Frost buys Richard Nixon

10 August, Beverly Hills, California

LIKE EVERYONE ELSE IN THE television industry, David Frost wasn't meant to be working in August. The eighth month was the dead time, a time of reruns and also-runs, when chat shows were hosted by stand-ins and everyone of importance was at the beach. Returning series had been scheduled, pilots already filmed, and even casting directors were on holiday. And as for the media? They were at the beach, too, although not such a salubrious beach, obviously. So, what was David Frost doing announcing a new show?

In Beverly Hills, on 10 August, Frost announced he had bought the exclusive rights to the 'television memoirs' of Richard Nixon. He said he and the former president had signed a thirteen-page contract the previous evening, exactly one year after Nixon had resigned the presidency.

Frost said there would be four ninety-minute programmes, which would be filmed immediately but would not be broadcast until after the elections of November 1976. He refused to disclose the amount of money Nixon would receive for the interviews, which would be filmed at the former president's estate in San Clemente, or when and on which stations the interviews would be televised. But details aside, it was still one hell of a scoop. At the time, Frost was commuting between London and Los Angeles on a regular basis and was one of the biggest names in television, on both sides of the Atlantic, so perhaps it was only natural that he had bagged Nixon. It was still a big deal though.

'I should make it clear that the former president has neither requested nor has he received any editorial control – whether in terms of the content or editing of the programmes, the use of newsreel footage or by way of prior knowledge of any of the questions,' Frost said. 'No subject, including Watergate, has been barred.'

Oooh.

The rights to the interviews were acquired by Frost on behalf of what he called an 'international consortium of broadcasting organisations'. Quite sensibly, he refused to identify any of the organisations. Irving 'Swifty' Lazar, the diminutive agent (he was five foot two) who was representing Nixon, had proposed to the networks the previous month that they buy the rights to a series of interviews with the former president. Both CBS and ABC rejected the proposals on the grounds that their company policies did not permit them to buy news exclusives. Even in 1975, this seemed terribly old-fashioned, while many simply didn't believe it. Surely everything in Hollywood was negotiable, right?

NBC, however, negotiated with Lazar on the premise that the interviews would be based on Nixon's written memoirs when they were eventually completed. 'Swifty' thought this was fine – while Lazar dubbed himself the 'Prince of Pitch', he had been given his other nickname by Humphrey Bogart back in the early fifties after Lazar had bet he could clinch five separate deals for the star before supper one day, and did. But Richard C. Wald, the president of NBC's news division, then said the negotiations had broken off after the network learned that Frost and others had been invited to enter into talks. At the time NBC had not even reached the question of editorial control over the interviews.

Frost said he'd already spent some time with Nixon in person and had found the former president 'physically well . . . He led us on a strenuous tour of the house so we could get an idea of filming locations, and I was astonished how vigorous he looked. It was a total contradiction of the ailing man I had pictured,' he said. Thirty-six-year-old Frost said the sixty-two-year-old Nixon had seemed 'totally in touch with reality'. Far more importantly for the interviews, he said at least a quarter would be devoted to Watergate. Frost said, without a hint of irony, 'I sense that he's ready to start reflecting on his life and his achievements. I have no reason to believe the ex-president will be less than candid.'

In a press release handed to reporters at a hasty news conference, Lazar was quoted as saying, 'From among the many people wanting to interview the former president, Mr Nixon chose David Frost because of Mr Frost's unique and wide-ranging experience.' Of course it had nothing to do with the money. When reporters saw the release, they smiled the crooked smile of every news reporter. Lazar would be parodied on The Muppet Show *as Fozzie Bear's agent, Irving Bizarre, who was so short he appeared only as a top hat atop a pair of shoes. Frost himself said he did not regard the paid interviews as chequebook journalism, mainly because Nixon was no longer in office. 'Each person still owns the rights to his own life after retirement.'*

The media was furious at Frost's exclusive, the New York Times *in particular. 'Since 1972, when* The David Frost Show *was cancelled, Mr Frost's interests have included stints as a BBC interviewer,' they wrote. 'Interest here in Mr Frost has centered on the celebrity status of his well-publicised romances, first with Diahann Carroll, the actress and singer, who broke the engagement and married someone else. Last year, the scenario was repeated when Karen Graham, a* Vogue *cover girl, married a Las Vegas hotel operator.'*

From George Orwell to Norman Rockwell

While David Bowie is rightly seen as the architect of everything he achieved – he was as expedient as he was clinical – occasionally he was the lucky recipient of serendipity. This is one such case, although it's also the record that rubber stamped the idea of him as a shapeshifter, a rock heretic who'd do anything for success. What he ended up with was perhaps his finest hour (or so), a work that wasn't 'plastic soul' at all (his words), but a song suite to rival anything by Marvin Gaye.

Young Americans by David Bowie

TOWARDS THE END OF 1974, towards the end of the sessions for a record that was still tentatively called *The Gouster*, Bowie was recording in New York's Record Plant, having moved from Sigma Sound in Philadelphia. He didn't have enough songs to finish the album, so Bowie quickly wrote 'Win', and then turned one of Luther Vandross's songs – 'Funky Music' – into 'Fascination' by changing the title and some of the lyrics. Vandross, who was singing backing vocals on the recordings, had also been performing his song.

'My friend Carlos [Alomar], whom I had grown up with, got a job playing guitar for David Bowie,' said Vandross. 'Carlos invited me to the studio. He and his wife, Robin, had gotten married a couple of years before and she is also a singer. As a matter of fact, Robin is one of the girls with whom I used to sing in the hallway. I stated making little vocal arrangements and showing them to Robin. I didn't know

205

that Bowie had overheard all this. He was sitting right behind me at the board, and he said, "That's a great idea. Put that down." So, I put it down and next thing you know one thing led to another, and I was doing the vocal arrangements for the whole album.' Vandross also started playing around with 'Funky Music' and, as soon as Bowie heard it, he said, 'I want to record that. Do you mind?' Vandross said, 'You're David Bowie, I live at home with my mother, you can do what you like.'

Bowie was trying to break America, on tour with *Diamond Dogs*, addicted to cocaine, and making a soul album. A new persona was beginning to take shape: a rail-thin, redheaded disco king in plaid suits and disco shoes. This, notionally, was the 'Gouster', a persona based on Chicagoan teenage black style in the sixties. However, in the end, Bowie opted to take on a musical costume rather than a physical one, finding his inspiration in the smooth sounds of Philadelphia soul. He was already spiking his live sets with covers of the Ohio Players' 'Here Today and Gone Tomorrow' and Eddie Floyd's 'Knock on Wood', but he was even more interested in what was happening in dance clubs. He had been toying with the idea for ages. 'When we were recording *Diamond Dogs* we worked on the song "1984", and he was already referencing Barry White,' said the producer, Ken Scott. 'He wanted the hi-hat and the strings to sound like they would on a Barry White record. He was already anticipating the sound of *Young Americans*. He had already moved on.'

Bowie had met the Puerto Rican session guitarist Carlos Alomar, who he wanted to work with Earl Slick. He then visited Sigma Sound Studios for the first time. Founded by songwriters Kenneth Gamble and Leon Huff, Sigma Sound was where Philadelphia International Records recorded the likes of the O'Jays, Patti LaBelle, Teddy Pendergrass and the house band MFSB. While he knew that he could only imitate the sound, he figured if he surrounded himself with the best musicians, then his stylised 'discophonic' brand of urban soul would be all the more genuine.

'My "Young American" was plastic, deliberately so, and it worked in a way I hadn't really expected, in as much that it really made me a star in America, which is the most ironic, ridiculous part of the equation,' said Bowie. 'Because while my invention was more plastic than anyone else's, it obviously had some resonance. Plastic soul for anyone who wants it. We really worked hard to make that record come alive.'

Alomar managed to get bassist Willie Weeks, who had already played with Stevie Wonder and Donny Hathaway, and drummer Andy Newmark, who had logged time with Sly and the Family Stone. The *Diamond Dogs* tour was still routed in rock arrangements, and as Bowie was spending his days making soul music, this was a period of transition, a hybrid. When the tour resumed, it had been radically altered and was nicknamed the *Philly Dogs* tour. It featured a medley of the Flares' 'Foot Stompin'' and the old jazz standard 'I Wish I Could Shimmy Like My Sister Kate'. As he made his way across America, the original musicians fell away one by one, until he was left with an almost entirely new line-up, many of whom had ties to R&B, funk and soul.

'Being a blues player, and having to make a living in New York City, I was already doing Sam and Dave, James Brown, Wilson Pickett and all that shit,' said Earl Slick. 'I was well-versed in all of that. So *Young Americans* wasn't a stretch for me. But he'd brought Carlos Alomar in, who was more from the pop R&B area. I was more of the Memphis shit. So the album was more of a Carlos thing, rightfully so, for what David needed. He was ruthless with people, especially musicians, and I was surprised I was kept. What happened was, when we finished *Diamond Dogs*, there was a break, and during that break, I was told that David had got another guitar player and that I was out of the picture. So I just went about my business. I was a little pissed off, because of the way it was done, and then out of nowhere I get a phone call: can you fly to LA, like now. So I was back. That was when we did a kind of a hybrid of the end of the *Diamond Dogs* tour going into the next phase, so this is September 1974. But by then Carlos was the MD so I was a guitar for hire. We went from doing a lot of stuff from *Ziggy*, *Aladdin Sane* and

Diamond Dogs, to doing versions that were less rock and roll, and went through a few different rhythm sections in the interim. We were also starting the *Young Americans* record, so it didn't feel that great. I have to be honest with you, it really didn't. Everything was disjointed. But *Young Americans* was a Carlos record, and he did a great job.'

Carlos Alomar was more concerned with Bowie's health, worried that he was so thin, and so delicate, that he might not be able to make the record. He was so worried he took him home to eat with his wife, Robin, in Queens.

'I offered to take him to the Apollo Theatre where I was playing with the Main Ingredient, as I also played in the house band there,' said Alomar. 'He'd always wanted to go and he loved it. He had this orange hair and his big, rimmed fedora hat and he was not only the only white person there, but he was also whiter than white. He is probably the whitest white guy I have ever seen. Imagine stepping out of a limousine looking like that and walking into the Apollo Theatre in front of a line of black people all lining up waiting to come in. He just strolls right in and gets a front row seat; he was in the lap of luxury. I introduced him to Richard Pryor that night, who couldn't make him out at all. He said, "What's this white dude doing in my dressing room?" David was fearless, he didn't care. I'm a Latin guy, Puerto Rican, although everyone thought I was black, so I took him to all the Latin clubs, and we hung out. He tried to hire me three times before he could afford me. So *Young Americans*, here we come! That album, self-labelled in his own right as "plastic soul". I defy that and say that it was a true, authentic soul record, he had the formula one hundred per cent. You'll hear a lot of people try and sing "Fame" and "Young Americans", but when it comes to the other songs on the record, they can't touch him. He had falsetto on top, and massive bottom on the bottom. I had been working with the O'Jays, the Ohio Players, so I knew. He knew exactly when he heard what he wanted, that Harold Melvin and the Blue Notes thing that he knew. He knew when he heard the Wes Montgomery guitar sound that he wanted, and when Luther started singing those background vocals, this became a real album. We did one song a day. Boom. Sometimes we did two songs a day, and David didn't even have the words written. He had to run home and stay up for four or five days just to match the moment.'

Keyboard player Mike Garson said that when they were recording the album, Bowie didn't get creative until two or three in the morning, when the cocaine arrived. 'At six in the morning he'd be wide-eyed and on top of things. It didn't matter what state he was in, he was always focused, always professional, always smiling. He relied a lot on David Sanborn and Luther Vandross on that record, because so much of the structures were complex, and the vocals were incredibly complicated.'

Bowie was chaotic, but focused; crazed because of his cocaine addiction, but channelling a ferocious amount of creativity. He was rich in ideas, bursting with energy and concepts for songs, overflowing with thoughts about every element of his stage sets, albums and costumes. His curiosity wasn't expedient, it was his defining characteristic. His decision to dump glam for soul was instinctual as much as anything.

'*Young Americans* was amazing because at that time David was one of the first white artists they'd recorded at Sigma Sound Studios in Philadelphia,' said Ava Cherry, who also sang on the album. 'David was immediately accepted by the black community. Before he arrived, I'd heard that some people were sniping about him coming to the city, but I never saw anything like that. I heard there were some players who didn't want to be on the record because David was white, but I don't believe that. David had already mapped exactly how he wanted to do it. He had met Carlos Alomar, and he didn't really care if there was a stigma. We went in there and just played away. David was very detailed about everything he wanted to do. He would be writing it down and he would write in a diary and write different things every day, all the time. He knew exactly what he wanted to do and was not ruffled by any of those things. He also did a show at the Tower Theatre in Philly, and that was a huge thing for the city. That's where he recorded *David Live*. It was great singing with Luther Vandross, as he was such a great singer, such an accomplished arranger, and our voices melded together like magic and butter.'

It was an incredibly romantic record and although Bowie was at pains to stress how insincere it was, it contains some of his warmest songs, with a vibe that wouldn't have been out of place on a Marvin Gaye album. 'Can You Hear Me?' for instance, is one of his greatest ballads, complete with David Sanborn playing the alto sax. Bowie knew it was good: in August, he told Anthony O'Grady, in an interview for the *NME*, '"Can You Hear Me?" was written for somebody but I'm not telling you who it is. That is a real love song. I kid you not.' Elsewhere, the widescreen aspirations of the title track feel like a disco Bob Dylan, the bubbly funk of 'Right' contains a line that followed Bowie throughout his career ('Never no turning back'), and the lolloping 'Somebody Up There Likes Me' was adapted from 'I Am Divine', a December 1973 outtake recorded by the Astronettes, Bowie's backing singers from the live performance TV programme *The 1980 Floor Show*. Even the songs that weren't especially romantic somehow felt sexy, although there is an innate anxiety fuelling the record, which feels like some kind of existential critique. It's not an overtly political record by any stretch, and yet it feels like social commentary. Bowie was a proper seventies' creation, despite having his first hit in the sixties, and would define it time and time again. This record proved he had real breadth, and a genuine ambition to escape the confines of the fame he had so desperately sought. Many thought it was odd that he wanted to make a soul record (at the time a lot of white rock critics tended not to like a lot of black music), and yet his ambition was easily as intriguing as the album's execution.

In early 1975, Tony Visconti was finishing production of the album and was due to fly back to London to mix it. On his last night in New York he got a call from Bowie telling him that John Lennon was coming to the studio that evening and, as he was a little nervous to be left alone with him, asked if he wouldn't mind coming along. Visconti was there in a flash. He had to ring the doorbell a dozen times – Lennon was also a little nervous, because he still hadn't had his green card yet and thought he might be the police. Bowie and Lennon were sketching pictures of each other, sounding each other out.

'I was there the day David brought John Lennon into the studio,' said Ava Cherry. 'He actually wrote a diary entry that day where he says, "30 January, introduced Ava to a Beatle." We were going in that day to record "Fame", and before the session David was freaking out because he was so nervous. He really admired John Lennon, and that day David was like a little kid. And then John comes in the door and John had those granny glasses on, right? And David looks and me and says, "He really does wear those granny glasses!" He really liked the fact that Lennon had the whole Lennon look. What you imagine John to be is exactly how he was like. Charming, funny, and they both hit it off immediately. They became really good friends. John was sitting there at one point with his twelve-string getting ready to play "Across the Universe", and he looks up and says, "Are we having a good time?" We were all so happy that John Lennon was so relaxed. David was just over the moon. He drew David a caricature of himself. And David put it in this solid gold frame. He really loved it. I didn't think "Fame" would turn out the way it did. I thought because John Lennon was on it that he was going to get lots of critical acclaim, but it was just a James Brown groove at one point.'

'I spent quite a lot of time getting to know Lennon, and I do remember we went to a lot of bars together,' said Bowie. 'We spent hours and hours discussing fame and what you had to do to get it, to get there. If I'm honest it was his fame we were discussing, because he was so much more famous than anyone who had been before. That period of my life is a little hazy, because I was heavily medicated! But I remember enough. I remember that Carlos and I were working on this riff, and I remember than it was John who started riffing on "Fame", screaming at the top of his voice in the studio. He was screaming, I was writing the lyrics, and Carlos was crashing through the riff. It all came together so quickly and so brilliantly. It was an incredibly intoxicating time and I can't quite believe that we didn't try and write more things together, because just being around him was breathtaking. He had all this energy, which I suppose I didn't expect when I first met him.'

A week after leaving for London, Visconti got a call from Bowie. He explained he'd gone back to the studio and recorded Lennon's "Across the Universe" for a lark and it turned out good enough to include on the album. He later played the track to Lennon, after which they decided to record a song together, which turned out to be "Fame". Bowie decided to include them on the new album, replacing 'Who Can I Be Now' and 'It's Gonna Be Me', much to Visconti's dismay.

The Old Grey Whistle Test's Bob Harris talked to Lennon about 'Fame' shortly after the release of the album. 'He said it developed from a simple riff, layering up and up in the studio, really building up from nothing. He said that as he was always in New York, and rarely left it, when the British guys came into town, they called him up and asked him to show them around. I remember John saying, "They don't need me, but it's nice to hang out." David had done the same thing, coercing Lennon into the studio to try and work on some tracks. "We were in the studio, and this riff started coming out, and we worked on it for three or four hours until the song was written." And it sounds like that, as it has that lovely loose spontaneity to it. And that was what John said: "It just sparked."'

Lennon also said that as well as the riff, he contributed some backwards piano, some falsetto, and a Stevie Wonder middle eight played backwards. 'And we made a record out of it, [and] he got his first No. 1.'

Psychologically, this was the start of David's very worst period, a time when his cocaine psychosis affected everything he did. He never looked less than extraordinary, though, and was always every inch the star. He was invited to the Academy Awards in 1975 and could not have looked stranger. He was wearing a huge Spanish black hat, a cape, and carrying a cane. His old friend Gus Dudgeon was there, the man who produced 'Space Oddity', and he couldn't believe his eyes: 'He looks fucking dramatic, like the ultimate gigolo. He's surrounded by eight tonnes of charisma, and everyone is gasping.' Dudgeon hadn't seen him for six years, but when Bowie saw him, he walked over, the mask dropped and he said, 'God, bloody hell, how are you?'

Dudgeon said, 'The Thin White Duke disappears and a completely different bloke appears. He gives us a hug and a kiss, sits and talks animated for fifteen minutes, and when he gets up to leave, on goes the hat, here comes the charisma and he's immediately the Thin White Duke again. What a star!'

A few weeks earlier, Bowie and Lennon found themselves backstage at the Grammys where Bowie had to present an award to Aretha Franklin. Before the show he'd been telling Lennon that he didn't think America really got what he did, that he was misunderstood. So, the big moment came, and he ripped open the envelope and announced, 'The winner is Aretha Franklin.' Aretha stepped forward, and with not so much as a glance in his direction, snatched the trophy out of his hands and said, 'Thank you. Wow. This is so good I could kiss David Bowie.' As Bowie slunk off stage, Lennon bounded over and gave him a theatrical kiss and a hug and said, 'See, Dave. America loves ya.'

'We pretty much got on like a house on fire after that,' said Bowie. 'He once famously described glam rock as just rock and roll with lipstick on. He was wrong of course, but it was very funny. Towards the end of the seventies, a group of us went off to Hong Kong on a holiday and John was in, sort of, house-husband mode and wanted to show Sean the world. And during one of our expeditions on the back streets a kid comes running up to him and says, 'Are you John Lennon?' And he said, 'No but I wish I had his money.' Which I promptly stole for myself . . . It's brilliant. It was such a wonderful thing to say. The kid said, "Oh, sorry. Of course you aren't," and ran off. I thought, This is the most effective device I've heard. I was back in New York a couple of months later in SoHo, downtown, and a voice pipes up in my ear, "Are you David Bowie?" And I said, "No, but I wish I had his money."

'"You lying bastard. You wish you had my money." It was John Lennon.'

Bowie was already developing a reputation for stealing; even though he was wildly original in terms of ideas, he was a professional magpie. This also applied to the cover of *Young Americans*. He'd seen the hand-tinted, forties-style Hollywood picture of the *Diamond*

Dogs tour choreographer Toni Basil on the cover of the September 1974 issue of *After Dark* magazine, and then commissioned the same photographer, Eric Stephen Jacobs, to shoot and airbrush an identical cover image for his own record. Bowie had originally approached Norman Rockwell, the celebrated American painter: 'His wife answered, and I said, "Hello, this is David Bowie," and so on. I asked if he could paint the cover. His wife said in this quavering, elderly voice, "I'm sorry, but Norman needs at least six months for his portraits." So, I had to pass.' *Young Americans* was such an important record, the first of the pivots that Bowie performed having become famous and the final sleeve has anyway become iconic – but imagine how more iconic it would have been had it been painted by the man responsible for the cover illustrations of everyday life for the *Saturday Evening Post* for fifty years.

Bowie always seemed obsessed with the adjacencies of fame. Two years after *Young Americans*, when Elvis Presley died in August 1977, Bowie briefly considered recording a tribute album, arranging classic Elvis songs for Iggy Pop to sing. He had been a fan all his life. Bowie's first performance, aged eleven, was an Elvis impersonation for an audience of boy scouts in Bromley. Years later, he would paint Elvis's 'TCB' ('Takin' care of business') lightning bolt logo onto his face on the cover of *Aladdin Sane*. His Ziggy Stardust concerts usually closed with the melodramatic 'Rock'n'Roll Suicide', which Bowie sang wearing an Elvis-style jumpsuit – copied by Bowie's designer friend Freddie Burretti from one of the King's – before departing the stage. This was immediately followed by the announcement, 'David Bowie has left the building.' In 1976, Bowie even tried to get Elvis to record 'Golden Years', only to have it turned down. It had been easy enough to get the demo to Elvis as they shared a record label, RCA, and bizarrely Elvis manager Colonel Parker thought it might be a good idea for the two stars to collaborate.

It's simple to see why Bowie pitched the song to Elvis. If you listen to his mid-seventies house band, it's remarkably easy to imagine them working with one of Bowie's classic Philly-soul records. They would have given it lots of hi-hat, wah-wah guitar and piping horns, employing the kind of Vegas swing that would have allowed Elvis to glide over the top, singing 'Come get up my baby . . . Come b-b-b-baby,' with all the baritone playfulness he could muster, just like he did on 'Teddy Bear' all those years ago. Bowie was so keen for Elvis to record the song he even sang a little like him on the verses, pitching his voice as close to the king's as he could. I'm not sure how Elvis would have coped with the falsetto breaks, but he would have done wonders with the growling parts. You can hear Bowie's Elvis on *Low* and *'Heroes',* as well as on 'Can You Hear Me' on *Young Americans* and on parts of 'Friday On My Mind' on *Pin Ups*. His 'Elvis voice' became something of a trope later on, too, and you can hear him delving down into the baritone on most of his post-*Never Let Me Down* albums, when Bowie was increasingly keen to reference his own past. For Bowie, Elvis was the consummate blueprint. Bowie himself has said that his debut album 'seemed to have its roots all over the place, in rock and vaudeville and music hall. I didn't know if I was Max Miller or Elvis Presley.'

When we first heard the 'Young Americans' single, as overly invested Bowiephiles, it was something of a shock, obviously, but in a way in brought Bowie closer towards us. Previously, he had been this alluringly strange figure, a kind of pan-sexual spaceman, as different from other pop stars as those other pop stars were from us. But overnight he had edged closer to his fan base, wearing expensive and designer versions of the clothes we wore on a Friday night. Of course, Bowie was influential, and of course he changed the way in which we all dressed, but he at least made it possible for us to try. And after all, we understood Oxford bags. Two years previously, when we tried to emulate the Ziggy Stardust look, it made fools out of many of us. With this new identity he at least gave us something of a chance. And it was soul music, right? We understood soul music, even if we were slightly bamboozled by Bowie's completely brilliant interpretation of it.

Like a Rolling Stone

26 August, 746 Brannon Street, San Francisco

ROLLING STONE'S *PUCKISH, RUMBUNCTIOUS EDITOR JANN WENNER wouldn't move the magazine's operations from San Francisco to New York until January 1977, but he started planning it in early 1975. He already had thirty staff members in the city, he was travelling there all the time (often it felt like a commute), and all the advertising money was on Madison Avenue. Of course, he could have moved everyone to Los Angeles, which was obviously the centre of the music industry (and where all the best cocaine came from) but he was moving for reasons of economy and efficiency. 'For the last three years we have had two main offices 3,000 miles apart, and this will allow us to consolidate our operation,' said Wenner. 'Furthermore, we believe in New York City.' They were going to take 20,000 square feet of office space in midtown, and he was excited. It wasn't good to have half your staff on the West Coast and half on the East. Why? Well, it was just too confusing. At the time, cocaine was still a vital ingredient in the exotic* Rolling Stone *diet. The office's photographic dark room reportedly doubled as a coke den called the 'Capri Lounge', which may have helped writers on deadline to work through the night. Star writer Hunter S. Thompson – whose entire output was fuelled by a cocktail of progressively contradictory drugs – wrote once that the whole thing was 'like being invited into a bonfire and finding out the fire is actually your friend.'*

Coke was the great social lubricant of the seventies, more so than sex, more so than music, money or booze. It was the great signifier, the *great leveller, the great convener. Wasn't particularly good for you, mind, but then that wasn't really the point of it. The point of it was to take it and then talk all night about why you'd taken it, and wouldn't it be a good idea right now to take a little more? If cocaine fuelled the music industry, then it stood to reason that it was probably going to fuel Rolling Stone, too. Los Angeles was awash with coke: it was almost as though it regularly blew in with the Santa Ana winds, stardust designed for the stars. Literally everyone was on it. Crosby, Stills, Nash & Young and their entourage were some of the city's worst offenders. Their 1974 tour was one of the most notoriously coke-fuelled tours of the decade. Insane amounts were being done, and the shows suffered as a result. The tour was largely confined to America, but it wrapped up at London's Wembley Stadium on 14 September. Stephen Stills was so far gone at this point that he reportedly believed that he had fought in Vietnam, signing autographs 'Stephen Stills, US Marine Corps'. Chicago, the jazz-pop band who might have given the impression they were above all this, were apparently so fond of cocaine that during a tour this year they put a fake phone booth on stage called the 'Snortitorium' so they could do lines mid-concert without going backstage. It was the kind of idea that occurred to people in the early hours of the morning, when cocaine ideas seemed most revolutionary. In any case, this particular idea was one that had probably occurred to most of the Top 50.*

Cocaine had such a stranglehold on the industry at the time, it almost acted as the coin of the realm. Not to be outdone, David Bowie moved to LA at the start of the year, renting a bungalow in Doheny Drive, opposite the Four Seasons in Beverly Hills. Ostensibly there to make a record, he became a professional cocaine addict, probably the best there had ever been.

'We were young blokes, doing what young blokes do,' said Black Sabbath's Tony Iommi, referring to their prodigious consumption of cocaine – one of their albums featured a track called 'Snowblind'. Holed up in a Bel Air mansion, their degenerate behaviour, involving substances and groupies, was, they said, worthy of Caligula. 'Nobody could control anyone else,' said Iommi, who almost overdosed one night

at the Hollywood Bowl. 'I was doing coke left, right and centre, and quaaludes, and God knows what else. We used to have [cocaine] flown in by private plane.'

One of the great benefits of being involved in the coke-flecked, quaalude-dropping rock and roll gestalt of the seventies was having access to the Starship, *the former United Airlines Boeing 720 that had been refurbished in the seventies as an airborne pleasure dome. The plane was used by the likes of Led Zeppelin, the Rolling Stones, Elton John, Peter Frampton and Alice Cooper. There was a bedroom with a king-sized waterbed, a drawing room with a fake fireplace, a thirty-foot brass-trimmed bar with built-in electric organ, a prehistoric video system stocked with everything from* Duck Soup *to* Deep Throat *and a brace of stewardesses to cater to the velvet-trousered minstrels fresh from their Hollywood Bowl gigs. The first time the Allman Brothers boarded they were greeted with 'Welcome Allman Brothers' rendered in lines of cocaine on the club room bar. Zeppelin's Robert Plant said his favourite memory of the plane was 'oral sex during turbulence', and Goldberg says the band's gargantuan manager, Peter Grant, would disappear with girls in tow to the bedroom and not reappear until the end of the flight.*

But Jann Wenner was keeping up. There is a story, never challenged by Wenner, that he once shocked a young reporter during an early profile by pouring out and then snorting a huge line of coke mid interview. According to Joe Hagan, the author of the contentious Wenner biography, Sticky Fingers *(a book described by Wenner as 'deeply flawed and tawdry' though he never sued over it), at a restaurant in New York, where Wenner was dining with Jackie Onassis, he and another* Rolling Stone *editor kept surreptitiously disappearing to the men's room. Halfway through an especially long anecdote, Wenner suffered a massive nosebleed, splashing his blood all over Onassis's clothes. Back at the* Rolling Stone *offices, he allegedly crowed to another editor, 'I bled all over Jackie – just like Jack!'*

The magazine's 26 August issue featured Garry Trudeau's cartoon extravaganza Doonesbury *on the cover.*

Nuclear Fusion

If Bob Dylan could go electric, then so could Miles Davis. Miles was such a genius that you often got the feeling he invented things just for the sake of it. Because he was bored. Because it was a Wednesday, and he hadn't stretched the vernacular of jazz for a good couple of days. While purists might point to Herbie Hancock and Frank Zappa as the true originators of fusion, it wasn't until Davis's incorporation of electric instrumentation on 1969's *In a Silent Way* that it properly became a thing.

No Mystery by Return to Forever

MILES DAVIS'S MUSIC ALWAYS PROVIDED pleasure, excitement, surprise, shock and energy; it also provided a roadmap of exploration. His audacious aesthetic of throwing something against the wall and seeing if it stuck resulted in the genuine development of jazz. *In a Silent Way*, which jump-started the seventies' fusion movement, marked the beginning of a whole new era. Synthesising an array of creative techniques and forward-facing musical ideas from rock, jazz, orchestral and electronic music, it created a blueprint. Recorded during a single three-hour session in July 1969 with producer Teo Macero, the album marked a decisive and definitive turn for both Davis and the future of jazz. Meditative, moody and minimal in approach, it was fundamentally something different. Some said that part of the mystique surrounding this album was that it seemingly came out of nowhere, but this is forgetting that Davis was perennially gifted with uncanny originality and visionary foresight. At the time, the great Lester Bangs said, 'This is the kind of album that gives you faith in the future of music.'

Bitches Brew, his electric bath of an album from 1970, was in some respects a funkier, more aggressive version of *In A Silent Way*. As one critic put it, his group got freakier, blacker and bloated with intensity. The guitarist John McLaughlin, who helped electrify Miles Davis's music, said it sounded like 'Picasso in sound'. It sold like hot cakes to rock fans.

'In my head, I'm saying – "Don't do that, do this,"' said Davis. 'But it happens so fast that when you listen to it again, you tell where you check yourself. If you jump on a horse and see he's on the wrong foot, you keep checking him until he gets to the fence – that's what I do when I'm playing. When I'm playing, I'm never through. It's unfinished. I like to find a place to leave for someone else to finish it. That's where the high comes in. If I know I left a perfectly good spot for someone else to come in – like, there it is! – then sometimes they don't come in. Sell out your plans! Especially when you put all that time in.'

It used to be said that jazz fusion is like Tabasco: it works in small doses. In 1975, however, when fusion was probably at its commercial peak, people loved it, loved it in a way they hadn't really liked anything before. Jazz nuts who wanted to embrace the world of rock, prog-heads who liked to tap their feet and critics who didn't really understand what was going on. Along with Weather Report, the Headhunters and the Mahavishnu Orchestra, Chick Corea and Stanley Clarke's Return to Forever were one of the core groups of the jazz-fusion movement of the early seventies. Corea formed the band in 1972, after playing on both *In a Silent Way* and *Bitches Brew*. He briefly formed an avant-garde jazz band called Circle with Dave Holland, Anthony Braxton and Barry Altschul, but after converting to Scientology, wanted to find a more accessible way to communicate with his audience.

Corea said that when Davis started to experiment, he became aware of rock bands and the energy and the different type of communication they had with audiences during a show. He'd see young people at rock concerts standing up to listen rather than sitting politely. They moved, too, something jazz audiences never did. Jazz audiences nodded, and not much else. It got him interested in communicating in the same way. People were standing because they were emotionally caught up in what they were hearing, and he related to that. He got fed up with the kind of audience he'd encountered at places like New York's Village Vanguard, for example. They drank quietly, they whispered surreptitiously and dispensed applause as though it were rationed. Jazz audiences didn't want bother anyone, didn't want to disturb the mood. So he started going to the Palladium and the Fillmore, rock venues where the crowds were noisier, more explosive, younger.

He knew that Davis didn't want to give up his form of jazz expression but also wanted to communicate with a younger, more emotional audience. So the sound and the rhythm of his music changed, which encouraged Corea to do the same. Which is how he came to set up Return to Forever, a band that to those who knew, already seemed like a supergroup: there was Stanley Clarke on bass, Al Di Meola on guitar and Lenny White on drums. At the time, Corea liked to call his music a 'hybrid'. It was appreciated as something much more by the people who enjoyed listening to it: the success of fusion musicians was based entirely on proficiency, skill, expertise and flair. Stagecraft didn't come into it, not sex appeal, not star power. This was a world where you didn't need to be an entertainer; all you needed was chops. In this respect, fusion was a lot like prog, although saliently far cooler, probably because of little more than reverse racism. (Q: What's the difference between a rock guitarist and a jazz guitarist? A: The rock guitarist plays three chords for a thousand people; the jazz guitarist plays a thousand chords for three people.)

'Artists have a very large responsibility,' said Chick Corea at the time. 'They can influence people because they're doing something people want to do – which is to create. The value of an art form is how much truth and honesty an artist communicates, so that he helps people to become more self-determined.'

While some rock artists had been gravitating towards jazz because they were bored and imagined that experimentation had no downside, many other jazz players started to dally with rock because they thought it might recapture their dwindling audience. And to a large extent, this worked. Of course, in the eyes of purists, fusion groups were starting to poison the sanctity of jazz, for reasons ranging from crass commercialisation of what was once a supposedly pure art form to what some thought was something more pernicious. David Rubinson, the co-producer of Herbie Hancock's *Head Hunters*, said that 'jazz fusion meant . . . white people playing black music.'

Fusion also meant embracing all the electric bells and whistles which were fuelling prog. The race towards the new technology was something that seemed primordial to those who were embracing the changes in jazz, often causing them to overuse and abuse these novelties (the Synclavier guitar effect for one, or the Clavinet), and yet the noise made by fusion bands was often so complex, and so complicated, that the sound they made was often more than the sum of its parts. Both for good and bad. Much of the early prog movement centred around Kent and particularly Canterbury, which spawned groups like Caravan and Soft Machine. They produced psychedelic records that included riffs played on flute and clarinet. By the early seventies, England had produced the likes of Nice, Gentle Giant, Gong, Van der Graaf Generator, Renaissance, Jethro Tull, Camel and Mike Oldfield, while embracing five-handed keyboard solos, flashy guitars, multi-part, half-hour classical suites with novelistic narratives, Mellotrons, symphonic experimentation, painstakingly intricate album covers, keyboardists in capes, Tolkienesque fantasies, travails from future days and scenes from a memory and an awful lot of noodling. Among the most enduring of the early prog bands were Yes (featuring the iconic keyboards of Rick Wakeman, drums of Bill Bruford and vocals of Jon Anderson), Emerson, Lake & Palmer (built around the epic suites composed by Keith Emerson), Genesis, Pink Floyd and Supertramp (led by co-songwriters Rick Davies and Roger Hodgson). Many of these groups started in a more traditional fashion before delving into odd time signatures, lyrics with literary allusions, and deliberately unusual chords. King Crimson, the brainchild of guitarist Robert Fripp, arrived almost fully formed as progressive rock artists on their debut album *In the Court of the Crimson King*.

None of them reached the noodly zenith of fusion, basically because fusion was almost never lumbered with vocals. There was actually some distance between the ornate, intricately arranged structures of prog and the freewheeling improv of fusion, although Frank Zappa was occasionally the man who bridged the two. Throughout his career, he had flipped between formats almost at will, jumping between prog, blues, modern classical, doo wop and whatever else was intriguing him at the time. But the merger of jazz and rock was one he exploited continually. That said, fusion, like jazz, saw the song as a framework for improvisation, whereas progressive rock, like rock, saw the song as an end in and of itself. And so it was in 1975.

Suspicion swirled around Zappa like smoke, as you were never quite sure if he meant what he was doing. He specialised in satire and was desperate to be acknowledged as the Grand Wazoo, the great oracle of sarcastic rock and yet he also wanted to be revered as a virtuoso, the noodly guy with a sense of humour. He made a lot of records, some of which were marvellous. For instance, as a soundtrack to the frenetic urban experience, few pieces of music could match 'Peaches En Regalia', the brilliant but incessant instrumental on his 1969 album, *Hot Rats* (it sounded like one big traffic jam, a bit like a non-linear companion piece to Jimi Hendrix's 'Crosstown Traffic'). Like Morrissey a decade later, Zappa would be remembered as much for his song and album titles as anything else: 'What's The Ugliest Part Of Your Body?', *Lumpy Gravy*, 'Don't Eat The Yellow Snow' and *We're Only In It For The Money*.

No Mystery was the fifth Return to Forever album, one of their most popular, and the one that garnered the most praise, winning a Grammy in 1976 for Best Jazz Performance by a Group (beating John Coltrane, Count Basie, Dizzy Gillespie and Supersax). It was a varied record; the first side consisted primarily of jazz-funk, while the second featured Corea's acoustic title track and a long composition with a strong Spanish influence. The hero track is 'Jungle Waterfall', a deep funk groove covered with sweeping wah-wah guitar and ostentatious bass; it had a terrific melody, but at absolutely no time did it allow the listener to ignore the technical wizardry that was swirling around them. The musicianship of Return to Forever was as important to the band and its punters as it was in Yes, Genesis, Emerson, Lake & Palmer or any of the other prog bands; if you weren't a real 'player' then you were bogus.

I saw these real players, when they came to London in 1976, playing the BBC Television Theatre in Shepherd's Bush in a specially recorded concert for Bob Harris's *The Old Grey Whistle Test*. It was an undiluted sixty-minute musofest. No sooner had they walked on stage than they all introduced each, using a solemnity usually reserved for introductions in Parliament or perhaps in court. First Chick Corea introduced Stanley Clarke, who introduced Lenny White, who then introduced Al Di Meola, who in turn then introduced Corea himself. There were no bogus men here, oh no: these were the real deal. And having told us just how important they all were, they started. Playing sometimes in unison and sometimes individually, the band incorporated the looseness of a jazz ensemble, the power-chord intensity of a rock unit and the synth-solo flash of a prog band (Return to Forever being just about the most prog-sounding group ever). In his Breton top, Corea alternated between synth and electric piano, playing a fluid solo one moment or producing a random special effect. Even when he was watching Di Meola or Clarke take their own solos (and, boy, did they love their solos), Corea looked like he was playing one himself. Watching from the circle, surrounded by hundreds of fusion nuts who were nothing if not rapt, I could sense that Corea was in a world of his own. At the same time, it was almost as if he couldn't quite believe so many people were attentively joining him and his band on their musical ride into a bright new tomorrow.

Fusion has, perhaps not so strangely, been the butt of many a joke: a couple go to see a marriage counsellor. They say their marriage is starting to collapse because they never speak to each other any more.

The counsellor tries to get them to talk, but they just sit there with their arms folded and their mouths closed. It feels like stalemate. So, he pulls out his electric bass and starts taking a solo, his fingers running up and down the fretboard as though he were Stanley Clarke in his prime, lost in reverie. Instantly, the couple turn to each other and start conversing for the first time in months. Shocked by this, the couple asks the counsellor: 'How did you know that would work?'

'Simple,' he says, 'everyone always talks during the bass solo.'

London Bombs

5 September, Hilton, Park Lane, London

THROUGHOUT THE DECADE, IN SOME respects, there was a siege mentality in London. Thirty years after German V1 bombs decimated parts of the city, the capital was once again the victim of vicious, random explosions. Irish republicans were quick to realise that one bomb in London was worth far more in terms of its impact and publicity than a dozen in Northern Ireland – bombing was acceptable there, but not 'here'. And so they took their fight to the UK in an effort to shake the resolve of the public and put pressure on the government. The city was not just a potent symbol with the national and international media, but it also housed a large Irish community in which IRA active service units were able to move easily. There was such an outcry over the number of IRA bombings in the city during the seventies that at one point The Times *came out in favour of a restoration of capital punishment.*

On 5 September, two people were left dead and sixty-three others were injured when the IRA bombed the lobby of the London Hilton in Park Lane. The police speculated that the bomb had been left under a marble table bearing vases of flowers. The Daily Mail *received a warning by telephone. One of the paper's switchboard operators said, 'A man with a soft Irish voice said, "A bomb will explode at the Hilton Hotel in ten minutes." I tried to hold him on but he merely repeated, "Hilton Hotel in ten minutes."'*

Having been notified, Scotland Yard immediately sent three officers to investigate, but they were not able to evacuate the building before the bomb exploded just after noon. A doctor in the casualty department of St George's Hospital at Hyde Park Corner, which the hotel overlooked, said, 'Some of the injuries were terrible. One chap is losing a leg, another had his ankle blown right off and is losing the bottom half of a leg.'

The Provisional IRA immediately claimed responsibility for the bombing, which was carried out by a unit called the 'Balcombe Street Gang'. It was the first in England since the Birmingham pub bombings ten months earlier, which had killed nineteen and wounded 180 more, and it marked the start of a renewed bombing campaign on the mainland. The group was responsible for a wave of more than twenty terror attacks in London before its members were arrested in December, after a six-day siege on the London street that gave them their name. They were also responsible for the assassination of Guinness Book of Records *co-author Ross McWhirter, who had publicly offered a reward for information leading to their arrest.*

By the time of the Hilton attack, the Troubles had already killed more than a thousand people on both sides of the Irish Sea, and the British public had become used to living under the threat of sporadic terrorism. In October, a second IRA bomb explosion outside Green Park tube station – just a couple of hundred yards away from the Hilton – killed one person and injured another twenty.

The IRA's activity left an unusual legacy in the capital – the removal of many of central London's rubbish bins, a favoured location for planting bombs. Most of the IRA's mainland attacks were carried out using explosive-packed vehicles. These car bombs were dumped near military targets, government buildings or notable public places. When planted in areas used by civilians, volunteers usually phoned through warnings, giving police some time to clear the area before the bomb detonated. These warnings were obviously not always effective, and often completely pointless. Some were not received until just minutes before detonation, while others (such as the Birmingham pub bombings in November 1974) failed to give exact locations. The IRA's use of telephone warnings also gave rise to thousands of hoax warnings, some made by the IRA itself.

It was usually impossible for authorities to know which warnings were genuine and which were false.

Even though the Troubles became part of the national psyche in the UK, they failed to inspire as much contemporary culture as you might have thought, especially in music. There were many plays, many documentaries and later films, but almost no music. No rebel songs. Nothing partisan, nothing expressing the futility or the waste. Oddly, the only two figures to address the topic directly were two ex-Beatles, John Lennon and Paul McCartney. Lennon's attempt at social commentary on the subject was the dirge-like 'Sunday Bloody Sunday', which sounded suspiciously like the Beatles' 'Come Together', and which was buried on his dreadful 1972 album, Some Time in New York City. *McCartney's was an atypical political nursery rhyme called 'Give Ireland Back to the Irish', which he disowned almost as soon as it was released. McCartney compounded the crime by releasing his version of 'Mary Had a Little Lamb' as a follow-up single. The only truly significant song about the Troubles at the time was 'The Town I Loved So Well' by the Dubliners, who were at least Irish. Van Morrison, who was already the industry's most prominent Irish rock star, steered away from the conflict almost completely. He occasionally made reference to the escalating troubles (in 'Saint Dominic's Preview' he referenced 'badges, flags and emblems'), but creatively only on his way somewhere else. Most of the time Morrison was either reminiscing about a more innocent time, recounting the sights and sounds of a bygone life, or escaping deep into his imagination, an oasis of romantic reverie. He was immersed in the country's culture and yet driven to escape it. The pictures he painted with his music were so impressionistic that they felt genuinely apart from anything resembling the present and were as far away from contemporary Northern Ireland as the music being made by Kraftwerk or Stevie Wonder. And he was completely unapologetic about it.*

Smells Like Downtown

In 1975, the streets below Manhattan's 14th Street were a no-go zone for anyone living above them. The Bowery – the home of CBGBs, where Patti Smith practised her art – was thought to be even worse. Smith's debut album practically screamed 'I am a bohemian'. The Velvet Underground and the Stooges had invented their genres just by falling out of bed and Smith was a keen student. What she learned was all fed into *Horses*. The poetry. The guitars. The androgyny. And the carefully art-directed black-and-white photography.

Horses by Patti Smith

THE ADVERTISEMENTS IN THE MUSIC press were adamant: 'Three-Chord Rock Merged with the Power of the Word'. The copy was laid across the image on the album cover, because it was the most powerful ammunition the record company had. With *Horses*, it was really all about the cover, at least it was at first, something approaching a work of art that arrived in the first week of November with almost no precedent, and even less fanfare.

The cover of *Horses* is considered to be one of the greatest of all time: a simple portrait of a woman in a crisp, oversized, white button-down shirt, a black ribbon draped over her shoulders, and dark jeans. A jacket is slung artfully over her shoulder. The wall she is leaning against is as white as her shirt, a blank canvas of sorts. Her hair? Not sure. Messy, definitely. The whole thing was carefully casual, so obviously meaning completely the opposite. Rock stars didn't look like this, weren't photographed like this, and wouldn't normally stare like this. Not in 1975, anyway. In Robert Mapplethorpe's stark, genuinely iconic portrait, Smith looked a little like a lot of things, but not a lot like anything we understood. Androgyny was the first thought, and a celebration of an attitude which was mildly confrontational but certainly attention-grabbing. She was reversing the female gaze, which at the time was rarely done (and when it was it was usually overtly sexual). She looked old and young at the same time (she was twenty-nine). To a lot of teenage boys used to staring at second-division rock stars grabbing their crotch and using microphone stands as penis substitutes, she looked quite scary. Smith says she had no idea what Mapplethorpe would do when she asked him to take her picture, but she knew it had to be 'true'. She had grown up in a semi-rural, hardscrabble working-class neighbourhood in southern New Jersey, across the Delaware River from Philadelphia, thirty minutes away by bus. So, everything you saw in Mapplethorpe's picture was an invention. Her mother was a waitress and her father worked the swing shift at a Honeywell plant, assembling thermostats. Blue collar creds for sure, with all the angst and ambition of the budding scribe hidden inside her.

When she started to be noticed, in New York 1974–5, her strikingly simple *androgynie* was surprisingly powerful, and it was copied by an entire generation of young boys and girls who wanted that sexy, Egon Schiele, tubercular look (we were more sheltered back then). This was the tortured brow as fait acompli. The rock and roll devotee as wannabe martyr. And it worked: critics in both London and New York were mesmerised by the cultural anomaly before them, anointing her immediately. For someone who aspired to be 'outside society', Smith was an astute social operator, at pains to act and look like the quintessential downtown beatnik, but always with her eye on the main chance. She was obsessed with both Mick Jagger and Jim Morrison, and it's the Doors singer she most resembled in terms of her schtick, sounding both young and old, while being fundamentally transgressive. Yet she transcended boho cultism by producing a hybrid, the sound of a post-beat poet, as she put it, 'dancing around to the simple rock and roll song'. The album was explicitly a New York record, written in the Chelsea Hotel and Smith's own MacDougal Street apartment then finessed at several of the city's storied venues, from St Mark's church to CBGBs.

Mapplethorpe's photograph was a downtown image, with downtown vibes. The picture was taken in the Greenwich Village apartment of Mapplethorpe's partner Sam Wagstaff, a picture that initially looked comprehensively urban. Mapplethorpe and Smith were already both veterans of a bohemian New York art scene that had created the fledgling ranks of the city's emerging punk fraternity, and *Horses* would immediately make her the North Star of everyone who came in her wake. What a debut it was, a statement of intent that didn't pull its punches. Just a while ago she had been a performing poet who had started to use guitars to amplify her stage craft, and now here she was, fronting a band who had just had their first album produced by the Velvet Underground's John Cale, signed by Arista Records for $700,000, with a slew of press interviews ahead of her.

To prepare for the session, Mapplethorpe told her to wear a white shirt, which he was keen to emphasise had to be clean (she bought it for the session from a Salvation Army store on the Bowery – the one she chose had RV embroidered on the breast pocket; she said it reminded her of *Barbarella*). She wore her favourite ribbon and her favourite jacket and he took about a dozen pictures, all in natural light. Embedded in the jacket is a horse pin given to her by friend, Blue Öyster Cult guitarist Allen Lanier. By the eighth one he said, 'I got it.' Smith asked him how he was so sure, and he just said, 'Trust me, I've got it.' And he had.

'[The cover] was a new way of looking, a new way of seeing that people didn't know how to express until then,' said the writer Richard Williams. 'It opened a door for people who were looking for a door, though they didn't know what it looked like they knew they would recognise it when they saw it.' Smith has described the album as 'my aural sword sheathed with Robert's image'.

'She was brunette, so she was obviously more serious than a blonde,' said Julie, a friend who had just turned ten at the time. 'Just like *Charlie's Angels* [which started the following year], you had Farrah Fawcett who was blonde, but Kate Jackson and Jaclyn Smith were different. They were the smart brunettes, especially Kate. But Patti Smith was cool. Dressed like a man, looked like she didn't care.'

As we all stared at the cover, thinking how studied it was, and how aligned with post-beat culture, a stronger impression wafted up: this was an image without vanity. I thought she was completely fabulous, and had obviously never seen an anyone like her, not on an album cover anyway.

By the time Mapplethorpe took this photograph, he and Smith had known each other for nearly a decade. They had been lovers until Mapplethorpe realised he was gay. Despite this, their relationship endured as artistic partners and friends. That intimacy is all over the picture on the front of *Horses*. You can read a lot into this picture, and later Smith would describe the way in which she built her own image – the ribbon being a reference to Baudelaire, the jacket over the shoulder a nod to fellow New Jerseyan Frank Sinatra, and the hair an homage to her sex obsession, Keith Richards. What the cover of *Horses* really was, was beyond gender – or post-gender, not that many at the time were grasping for that. She is making no concession to the consumer, no overture to the photographer. Five years later she may have been giving them the finger, but not now, not in 1975, when punk rebellion was still seen as a small time, almost retro thing. Critics often say that Smith was a harbinger of punk, whereas she probably felt she was the last in a line of boho renegades. The last, not the first.

She was also channelling Bob Dylan, another of her heroes. When she moved to Greenwich Village at the age of seventeen in 1964, she once saw him coming out of his apartment. She felt like she'd arrived. Eleven years later, in the summer of 1975, she met her idol for the first time after she performed at the Bitter End, the club on Bleecker Street. 'Backstage after the show, somebody taps me on the shoulder and says, "Any poets around here?" I turn around and it's Dylan. Unbelievable! A few days later we ran into each other on the street and Bob pulls out a page from the *Village Voice* and it's a big photograph of me and him backstage. And he goes, "Do you know these people?"' She really had arrived this time.

'She does look like a dandy in the photo – I think it's the way the jacket is slung over her shoulder,' said the historian Cally Blackman in the *Guardian*. 'And of course she looks like a boy or a man, so perhaps she is also continuing in the well-established tradition since the late seventeenth century of women who appropriate male dress for whatever reason, to express sexuality or the attempt to reform dress as stage costume, or merely for comfort.'

She was simply saying, this is me, if you know me at all, which you don't. In lesser hands, with a lesser talent, the cover of *Horses* would have looked very different. First of all, Smith would have been a man, he would have had a packet of cigarettes (Camels, maybe, or Marlboro Red) pushed high into his rolled-up chambray shirt sleeve, and he would have been looking alluringly at the camera, with a hint of a sneer. There would have been muscles. He would have been leaning against a vintage car, with the sunset conveniently behind him. He would have been wearing jeans, scuffed motorcycle boots, and a longish greasy quiff. Who knows, he may even have been smiling. The photograph would have been in colour, and the record would not have been called *Horses*. More likely it would have been called *Troubadour*, *Outlaw* or *Cars*. May have even been called *Jersey Boy*. It would have looked down-home, it would have had suggestive and awkward messianic overtones, and it wouldn't have been very good.

As far as the world knew, Patti Smith existed only in black-and-white. Her album cover was monochrome, and all the pictures we saw in the music ideas were harsh black-and-white photos of her and her band either onstage or framed suitably against some form of urban deprivation. She hadn't broken into mainstream media yet, which was led by colour, so she remained a marginal performer, performing in a world that deliberately felt cheap and alt. And it worked.

Mapplethorpe's photo is both high-art European and high-fashion gloss, but it certainly wasn't glamorous enough for Clive Davis, the founder and president of Arista records. He was appalled by the image and begged Smith to reshoot it. He objected to her messy hair, the lack of make-up and her man's tie. He was also, reportedly, critical of the slim trace of facial hair on Smith's upper lip. But Smith's contract with the label afforded her complete artistic control, and she rejected all his advice, including the suggestion the art department airbrush her 'moustache'. 'I felt it would be like having plastic surgery or something,' she said. 'I told them, "Robert Mapplethorpe is an artist, and he doesn't let anyone touch his pictures." I didn't know that for sure, maybe he wouldn't have minded, but I would have.' Of course she would have; after all, she was a poet masquerading as a rock star, unlike all the rocks pretending to be poets. There may have been elements of her work that seemed pretentious, but the very last thing she was doing was pretending.

The cover worked perfectly. Viv Albertine, who was soon to form the Slits, said she had never seen a girl who looked like this. She said she was her soul made visible; all the things she hid deep inside herself that couldn't come out. It was her natural, confident look as much as anything that intrigued her. She didn't want to copy her style, but it gave her the confidence to express herself in her own way. There would be hundreds of thousands like her, girls who didn't want to be boys, but who absolutely wanted their agency. With *Horses*, they got it, without even having heard a word of it.

Boys loved it too, as it was an instant outlier. If, in the early seventies, the simplest way to display your progressive leanings was by carrying around an album cover, preferably while you were wearing a greatcoat, regardless of the month (the most popular choices at the time appeared to be *The Dark Side of the Moon* and *Led Zeppelin II*), by 1975 it had become such a cliche that it no longer had the appropriate power. If you carried an album cover through town, under your arm, with the front of the sleeve facing outwards, almost begging for approval, you just looked stupid. So I remember thinking how odd it was that I saw a boy I knew carrying *Horses* with great pride through Battersea Park. He was a little odd, although appealingly so; he even dressed a little like Patti Smith on the cover. A year later he would have probably been dressing like Joe Strummer or Pete Shelley (we lost contact, so I don't know), but his wholesale appropriation of *Horses* was semaphoring the fact he was outside of society.

What was inside the sleeve was equally as transgressive, equally as bold. There are few better side one, track one songs than Smith's interpretation of Van Morrison's 'Gloria', an incendiary piece of beat rock and roll that starts with a line from one of her early poems: 'Jesus died for somebody's sins, but not mine.' It was avant-garde art punk – three-chord rock liberally drenched in the contrary and the arch. You could practically smell the well-thumbed William Burroughs paperbacks. *Horses* positioned Smith as 'punk's poet laureate', a woman who could articulate the power of rock and roll in a way that contextualised its ambition and its transgression. 'She belonged to a time,' says Richard Williams, 'but she didn't belong to a movement. She existed slightly to one side.'

The earliest theory of art, espoused by the Greek philosophers, proposed that art was mimesis, the imitation of reality. Patti Smith's *Horses* was very much an imitation of reality. Never had there been a record that screamed 'I am a bohemian' so loudly, so plaintively. While the Velvet Underground, the MC5 and the Stooges had invented their own particular genres seemingly just by falling out of bed, Smith was a keen student of everything alternative. And what she had learned was all fed into *Horses*. The poetry. The guitars. The androgyny. The death wish. This was the tortured brow as fait accompli. The rock and roll devotee as wannabe martyr. And it worked: critics in both London and New York were mesmerised by the cultural anomaly before them, anointing her immediately. For someone who aspired to be 'outside society', Smith was an astute social operator, at pains to act and look like the quintessential downtown beatnik. 'Patti wanted to look like Keith Richards, smoke like Jeanne Moreau, walk like Bob Dylan and write like Arthur Rimbaud,' said one of her pre-fame flatmates.

Of course, in the studio that depth and maturity was steered expertly by John Cale. When he was originally asked to produce the record, he initially didn't know if he was being asked out on a date or whether he was being hired for a job of work. But Smith was incredibly direct, telling Cale precisely what she wanted. He in turn says that the experience was intoxicating, and although he never knew where the record was going to fly off to, it was 'miraculous in its flight'. The record is a masterpiece, much of which is down to Cale.

'No one expected me,' said Smith at the time, by way of explanation. 'Everything awaited me.' And, beyond cocky, and convinced of her own importance, she meant it. 'My picking John was about as arbitrary as picking Rimbaud,' Smith told the journalist Dave Marsh a year later. 'I saw the cover of *Illuminations* with Rimbaud's face, y'know, he looked so cool, just like Bob Dylan. So Rimbaud became my favourite poet. I looked at the cover of [Cale's album] *Fear* and I said, "Now there's a set of cheekbones." In my mind I picked him because his records sounded good. But I hired the wrong guy. All I was really looking for was a technical person. Instead, I got a total maniac artist. I went to pick out an expensive watercolour painting and instead I got a mirror. It was really like *A Season in Hell* [also by Rimbaud], for both of us. But inspiration doesn't always have to be someone sending me half-a-dozen American Beauty roses. There's a lotta inspiration going on between the murderer and the victim. And he had me so nuts I wound up doing this nine-minute cut ['Birdland'] that transcended anything I ever did before.'

Charles Shaar Murray was an *NME* star, and it was he who got to review *Horses*. And he loved it. Not only did he say that first albums this good were pretty damn few and far between, but he also said it was better than the first Roxy album, better than the first Beatles and Stones albums, better than Dylan's first album, as good as the Doors and Who and Hendrix and Velvet Underground albums.

CSM (as he was known colloquially in the pages of the *NME*) was, along with Nick Kent, probably the paper's most famous writer, and as well as being on good terms with David Bowie (he had broken the Ziggy retirement story), and revered by many of the bigger acts of the time, was also known to be a terrible cynic (Kent and Shaar Murray were so famous that they were regularly asked for autographs at gigs; according to Patti Smith's guitarist Ivan Kral, Smith had her eye on Kent, although the relationship was never consummated). Shaar Murray was averse to hype, especially surrounding new, leather-jacketed, post-New York Dolls types or anyone coming out of the CBGB world. But he positively loved the Patti Smith record, and said it was hard to think of any other rock artist of recent years who arrived in the studios to make their first major recordings with their work developed to such a depth and level of maturity.

He used his review to chastise her for a) being a poet, b) being from New York, and c) rather ungallantly, not being stereotypically attractive. Yet he also said *Horses* was some kind of definitive essay on the American night as a state of mind, an emergence from the dark undercurrent of American rock that spewed up Jim Morrison, Lou Reed and Dylan's best work. In full flight he said the record was night-wailing, street-corner blues, the midnight flight out of Gasoline Alley, to Desolation Row, a thrashing exorcism of public and private demons. 'Patti Smith's album hips you to just what's wrong with a lot of other stuff you've been listening to, tips you off as to who's really doing it and who's just going through the motions,' he said. '*Horses* is an album in a thousand. I'm not gonna jive you about how influential it's going to be (in terms of it stimulating dozens of toy Patti Smiths to come crawling out of the woodwork I hope it has no influence at all), but, God knows, it's an important album in terms of what rock can encompass without losing its identity as a musical form, in that it introduces an artist of greater vision than has been seen in rock for far too long. The fact that Patti Smith is a woman may well alienate listeners who are prepared to be receptive to a basically passive female intelligence (like Joni Mitchell), but may find an album of extrovert, ferocious female intelligence (like this one) somewhat unnerving. Not to mention the fact that people always get weird in the presence of a powerful sexuality expressed by someone who they may not happen to find attractive. However, I'll say it again . . . first albums this good are pretty damn few and far between.'

We believed him. It's no exaggeration to say that Shaar Murray's review made Patti Smith a star. Jonh Ingham gave it a five-star review in *Sounds*, but *Melody Maker*'s Steve Lake attacked the album as an embodiment of 'precisely what's wrong with rock and roll right now', panning it as 'completely contrived amateurism' with a 'so bad it's good aesthetic'. It seemed as though battle lines had been drawn, demarcations that would be applied in a far more stringent and hysterical way during punk.

Smith said her vision for her debut album was 'to make a record that would make a certain type of person not feel alone. People who were like me, different. I wasn't targeting the whole world. I wasn't trying to make a hit record. Psychologically, somewhere in our hearts, we were all screwed up because those people died. We all had to pull ourselves together. To me, that's why our record's called *Horses*. We had to pull the reins on ourselves to recharge ourselves. We've gotten ourselves back together. It's time to let the horses loose again. We're ready to start moving again.'

Did it all feel a bit marginal? A bit NY specific? A bit narcissistic? Even eschewing vanity, she was still staking a claim. But then what Smith was creating was less an artefact and more a socio-economic type. Before *Horses*, people didn't really look like Patti Smith. After it was released, her look – her mien – became an archetype, a Camden Town/SoHo/East Berlin/Left Bank/Venice Beach prototype as popular as it was easy to replicate. It's sometimes difficult to look at its cover and wonder what all the fuss was about, but after it was released at the end of 1975, the world would never really be the same. Everything was fresh: to paraphrase Susan Sontag, it wasn't simply that the sixties had been repudiated, the dissident spirit quashed, and made the object of intense nostalgia, it was the fact Smith was reframing 'beat'. She may have been accidentally spearheading the Blank Generation, but in reality, she was a feminist iteration of the Beat Generation. She would go on to be called the 'princess of piss', 'the keeper of the phlegm' and 'the wild mustang of rock and roll', but she was never any of these things. In hindsight she may have been the link between Iggy

Pop and Television, between the MC5 and Talking Heads, but what she really was, was a downtown poet who accessorised her work with scratchy guitars, turning herself into an androgynous independent. Women in 1975 just didn't look like Patti Smith, and they didn't sound like her either.

'Basil!'

19 September, Wooburn Grange Country Club, Wooburn Green, Buckinghamshire

HAVING JUST REALISED, YET AGAIN, that he had inadvertently made a massive faux pas, probably one obvious to everyone in the dining room of his low-rent, lower-middle-class Torquay hotel, Basil Fawlty would give one of his harrowing grins, as if he had just received a shock through alligator clips attached to his genitals. Sybil, his wife, would smile indiscriminately and rather obliviously at anyone around her, as though nothing at all had happened, before laughing like a deranged hyena. Yes, welcome to Fawlty Towers, *the worst hotel in the world and, even at launch, acknowledged as one of the funniest sitcoms ever broadcast.*

After the monumental success of Monty Python's Flying Circus, *which ran from 1969 to 1974, revered the world over for its surreal and anarchic sensibility, it seemed strange that John Cleese would want to follow it with a traditional sitcom, let alone such a formulaic farce. Yet the BBC's* Fawlty Towers *was just as surreal, and certainly just as funny. It was genuinely different. Mervyn Rothstein of the* New York Times *called it the 'paragon of sitcoms', and reviewer and author Clive James recalled 'retching with laughter' when he watched it. At the time, there were only three TV channels in the UK, so a hit had enormous cut-through, while popular programmes were regularly watched by between 25 and 50 per cent of the population. The show, which was broadcast at 9 p.m. on BBC2, was sandwiched between* The Money Programme *('Taxed Beyond Belief: How much of your pay goes in income tax? Today the lowest rate of income tax is thirty-five pence in the pound, and for some people it is as high as ninety-eight pence')* and Leap in the Dark: 1 – The Rosenheim Poltergeist *('Colin Wilson introduces four cases of extrasensory perception').*

Fawlty Towers *was a beautifully observed examination of petit bourgeois values, manifesting themselves in the prosperous form of the misanthropic arch snob Basil Fawlty, played by Cleese. It was light years away from the soon-to-be-launched* Saturday Night Live, *as it focused on the British inability to reconsider what life had prepared for them. In a way it was a celebration of the way in which Britain had become a global force almost in spite of itself. Fawlty was beautifully drawn, a man like Tony Hancock, Harold Steptoe and Arthur Lowe's Captain Mainwaring, who all nursed frustrated social aspirations. Cleese's creation was a rude, long-suffering and monumentally unreasonable hotelier, while Prunella Scales played his martinet of a wife, Sybil; they were supported by Andrew Sachs as Manuel, a dim-witted and incompetent Spanish waiter, who frequently found himself on the receiving end of Fawlty's temper, and Polly, a calm, pretty waitress played by Cleese's then-wife, Connie Booth, who co-wrote the show with him.*

As Monty Python was so revered by Americans, there was an assumption that Cleese was a legitimate comedy export legend, like Peter Cook or Dudley Moore, a gentleman maverick; if he wasn't, then Fawlty Towers *made him properly famous. The show may have reinforced the idea that Britain was the funniest nation on Earth, but it also reminded us and everyone else that we were actually the most conservative. It is still regularly considered to be the finest British sitcom ever. The British Film Institute even voted it the finest British television programme ever, full stop (above both* Cathy Come Home *and* Doctor Who*). One critic went so far as to say it was, after the NHS, arguably Britain's greatest achievement since the Second World War. It's certainly lasted better.*

Each episode was an intricately choreographed dance of escalating tension building to an absurdist crescendo. The final episode of the first series (a further six would be aired in 1979) was called 'The Germans' and was a stark reminder that we were only thirty years away from the end of the war. While suffering the effects of a concussion, Fawlty tends to a party of hotel guests from West Germany. Despite telling his staff, 'They're Germans! Don't mention the war!' he keeps ignoring his own advice. A litany of anti-German comments culminates in Fawlty (who by episode six had morphed completely into Cleese) goose-stepping around while shouting, 'Don't mention the war!' Hurt and bewildered, the Germans are left wondering aloud how such an idiot as Basil could ever have beaten their ancestors in two world wars. At one point, Fawlty says, 'Ah, wonderful! Wunderbar! Ahh! Please allow me to introduce myself, I am the owner of Fawlty Towers. And may I welcome your war ... your war ... you wall ... you all ... you all, and hope that your stay will be a happy one. Now, would you like to eat first or would you like a drink before the war! Er ... trespassers will be tied up with piano wire ... Sorry, sorry!'

In 2020, the BBC momentarily took this episode off its steaming service because of what it called 'racist slurs'. Cleese, who had made a habit of calling out what he considered to be hypocritical behaviour, said, 'The BBC is now run by a mixture of marketing people and petty bureaucrats. I would have hoped that someone at the BBC would understand that there are two ways of making fun of human behaviour. One is to attack it directly. The other is to have someone who is patently a figure of fun, speak up on behalf of that behaviour.' The episode was soon reinstated, with warnings highlighting 'potentially offensive content and language' accompanying it.

Unlike most of Cleese's comic creatures, which were invariably either based on stereotypes (and in some cases creating them), Fawlty was based on a real-life hotelier, Donald Sinclair, who ran the Gleneagles hotel in Torquay. In the early seventies, the cast of Monty Python had stayed at the hotel while they were filming in Paignton, nearby. Cleese was fascinated by Sinclair's appalling behaviour, which included criticising fellow Python Terry Gilliam's 'American' table etiquette and throwing Eric Idle's briefcase out of a window 'in case it contained a bomb'. Cleese described Sinclair as 'the most marvellously rude man I've ever met'.

How to Talk Dirty
and Influence People

Feminism came in many forms in the seventies, but the movement didn't know what hit it when it discovered Millie Jackson. She was called 'the queen of raunchy soul', 'the godmother of rap' and 'soul's queen of sexual outrage', because of her signature, no-holds-barred lyrical content and her long raps – profanity-laced, sexually explicit stories and jokes – interwoven through her songs and live sets. The *Washington Post* would call her a 'virtuoso of vulgarity', oblivious to the fact that this was a persona she had honed to perfection.

Still Caught Up by Millie Jackson

ONE UNUSUALLY COLD THURSDAY NIGHT in 1964, when she was barely twenty, Millie Jackson was sitting with some friends at the Psalms Café on 125th Street in Harlem. The restaurant held an open mic on Thursdays, and Jackson was monstering a young singer who had the made the mistake of thinking the stage was a safe space. Her friends bet her five dollars to get up herself and sing, and so, after some cajoling (actually not much), she did – even though she had no training as a singer. A club promoter in the audience offered her a gig the following week, and the rest is New York history. She got a manager, a record deal, and a reputation in a New York minute. She hustled throughout the sixties, having the occasional minor hit, but it wasn't until the seventies she began to get real traction.

One of her biggest early songs was 1973's 'It Hurts So Good', which appeared on the soundtrack to the blaxploitation hit *Cleopatra Jones* (commissioned after the success of Isaac Hayes's *Shaft* (1971), it was also the first blaxploitation film to use martial arts as part of its promotion). Sick of being compared to Gladys Knight (their singing voices were almost identical), Jackson decided to change the content of her songs. Sick of hanging around in a safe place, she decided to get personal. Uber personal. The result was *Caught Up*, Jackson's triumphant and revolutionary 1974 concept album about cheating, with the first side featuring songs from the perspective of the mistress and the second side from the wife. The concept was simple: the soul story of a love triangle, narrated by two women cojoined with the one man.

While Curtis Mayfield and Marvin Gaye were creating cohesive bodies of work that reflected community, racial and environmental turmoil, Jackson focused on what was happening in her own home, in her own bed. This was a world away from Rick Wakeman's rather parochial albums about Jules Verne, Henry VIII or the myths and legends of King Arthur. This was a deeply personal, passionate mini-soap opera. 'We knew we were onto something, after recording a [nine-minute] version of "(If Loving You Is Wrong), I Don't Want to be Right",' the first song on *Caught Up*, Jackson explained in an interview. 'Then somebody in the studio asked, "What now?" And I said, "We finish the story. We've heard from the girlfriend, but what about the wife?"' *Caught Up* was filthy, too. Long before contemporary rap albums caried parental advisory warnings, Millie Jackson's sexually explicit soul records bore the admonishment, 'For Mature Audiences Only' ('Contains explicit language which may be considered objectionable by some listeners'), which obviously added to their broader appeal.

While her formative albums came out years before the beginnings of hip-hop, the genre eventually drew on Jackson for influence, as her spoken-word style and caustic, don't-fuck-with-me energy laid the groundwork for decades' worth of female rappers. Talking trash, looking you straight in the eye, and waiting for a response. Her mid-song, minutes-long tales of heartache and betrayal, usually delivered in typically frank and profane language, gave Jackson a cult following for her originality, but also made her records all but unplayable on the radio. Not only were her songs full of profanity, they also segued into each other. Radio DJs mostly tended not to play her records, as they never knew when each song would end.

Her record company were full of praise but, while they thought the album was 'gorgeous and wonderful' (according to one executive), who the hell was going to play it? 'It was the everyday concept of a love triangle,' said the journalist Suzanne Moore, who loved *Caught Up* when it was released. 'Jackson tells us how it is to be the mistress, how it is to be the wife, how it is to know a relationship is over, how desire waxes and wanes as domesticity takes over, how to leave a lover.' The record was modern, shocking, and mature. And made by a woman. Like no other black female singer of the seventies, she played a crucial role in representing and empowering her communities. Through her music, she addressed social issues as well as sexual mores, celebrating heritage and pride, and becoming a symbol of strength and resilience in the process. Her experimentation expanded the boundaries of what was considered traditional music for black female artists at the time. Artists like Diana Ross, Gladys Knight, Aretha Franklin and Chaka Khan were becoming synonymous with musical excellence and cultural significance, but Millie Jackson was a trailblazer. A potty-mouthed one at that. She became an agent of social change by focusing on the minutiae of relationships and by bringing the bedroom into the foreground – not as a sexual fantasy but an unavoidable battlefield. Although, because she was speaking from experience, Jackson always understood the importance of compromise as well as perseverance. And she knew exactly who she was singing for. Her audience was primarily black, female and working class. As she once told *Atlanta Magazine*, 'I didn't sell record to bougies. It was the poor people who bought my music. The women who bought Diana Ross did not buy Millie Jackson. The people in the projects understood me. I was down and dirty. I told you like it was.'

Jackson's vocal style had an intimacy and an easy familiarity that made the listener feel like they were in a private conversation.

In that sense it was almost like a bespoke recording; her schtick made her live performances communal, and her records entirely particular. She was like the best friend who gave you the truth you didn't want to hear, laden with pragmatic common sense.

'One side of the entire album is about the girl going with the married man,' Jackson said. 'But the second side of the album, I thought the wife should have her say. So, it's from the side of the wife and what she thinks about being cheated on. Her confrontation with the girlfriend. You know, "All you're getting' is my leftovers, digging out of love I done picked over. You oughta leave my man alone, find one of your own." When I write a story like that, I like to balance it out so people on both sides can see what's going on.'

Recorded down in Muscle Shoals, Alabama, Jackson, along with producer Brad Shapiro and house band the Muscle Shoals Swampers, created a classic that would land the Thomson, Georgia, native the highest-charting album of her career and a Grammy nomination. And it wasn't just the critics who loved the album, so did the record-buying public. They helped it to reach No. 21 in the US Billboard 200 and No. 4 in the US R&B charts. The album went gold in the States within six weeks of release. Jackson immediately found herself becoming better known for her recorded 'raps' than for her singing. It was something she was keen to repeat.

Consequently, a year later, in July 1975, there was a sequel, *Still Caught Up*, which reprised its themes of adultery and recrimination. Where side A of *Caught Up* featured Jackson singing from the mistress's point of view and side B from the jilted wife's point of view, *Still Caught Up* begins with the wife on side A and concludes with the mistress. Some sequel: this was the *Godfather II* of soul albums, a recording of political nuance, anger, and a dispassionate eye that some thought was better than *Caught Up*. Her trademark 'rapping' – the long intros, interludes and dialogue breaks Jackson skilfully blended into her songs – was also an accident. She had no formal vocal training, so she was not a strong singer at the beginning of her career. When audiences in supper clubs and lounges started talking to each other and turning their attention away from the performance, she started talking to them to keep them engaged. Nina Simone used to do this, but Simone was aggressive (this was a period in which her music was becoming increasingly militant and less commercial). Jackson wasn't. Her approach was almost consensual. This became a key part of her artistry. Jackson didn't just sing you a song, she told you a story, the whole story. No one had done this before in pop, certainly not a woman. Adultery and torrid affairs certainly weren't new topics in music, but they were relatively new to R&B. 'Infidelity is my whole repertoire,' she said once. The primary topic of Jackson's music, after infidelity, was sex. Not making love. But sex.

The first song on *Still Caught Up* is a live rendition of Tom Jans's 'Loving Arms', which not only showed off Jackson's extraordinary signing voice, but also reconfirmed the intensity of her stage performance. In fact, the whole album could have been recorded live. It gave an indication of Jackson's enormous convening ability, and the way in which her shows were becoming more than performances. They felt cathartic for everyone in the room. She was now a 'personality', a female Richard Pryor, a black Joan Rivers, a supercharged Mae West, performing material that would have made even Bette Midler blush. Her introductions were nothing if not adulatory: 'So now ladies and gentlemen it is star time … ARE YOU READY FOR STAR TIME? Thank you and thank you very kindly. It is indeed a great pleasure to present to you at this time, the HARDest-working woman in show business … The Amazing Miss "'Please, please, please,' my ass, I DEMAND" herself … THE STAR OF THE SHOW … MILLIE JACKSON AND HER FABULOUS SOUL REVUE!!' It was aggrandising, but it wasn't selling false goods. And she was a woman. As Richard Williams wrote in *The Times*, reviewing one of her shows at the time, 'One is so accustomed to the sexual boastfulness of male rock and soul singers that it comes as an instructive change to witness the same pose enacted by a woman. Utterly eschewing the submissiveness of a Diana Ross or the downtrodden despair of an Aretha Franklin, Millie

Jackson is possibly the first since Ma Rainey and the early classic blues singers to present herself as a rapacious sexual conqueror.'

The honesty was searing. Whether on stage or on record, the way she channelled both wife and mistress was pitch perfect. She almost casually described a typical relationship, of a married man coming over two or three times a week, giving 'a little bit'. According to the mistress, she was already 'two up on the wife', because after you're married, 'you only gonna get it once a week'. On payday, says the mistress, her man can come over and 'give you a little bread too, and I like that'. But the sweetest thing of all? When she went to the Laundromat, she didn't have to wash his funky drawers.

Infidelity and sexual demands from the woman's point of view was topical fare for dirty blues, not R&B. Up until Jackson, raunchy soul had largely been the domain of male singers. This hadn't been the case with dirty blues, where the likes of Bessie Smith, Lil Johnson, Lucille Bogan and Wynonie Harris pumped out such risqué tunes as 'It's Tight Like That', 'You Never Miss Your Jelly Till Your Jelly Roller is Gone', 'Need a Little Sugar in My Bowl' and 'King Size Papa' (eye-poppingly filthy, with a giant neon *wink* and nary a four-letter word). Many were very explicit rather than just full of innuendo. The most extreme examples were rarely recorded at all, a notable exception being Lucille Bogan's obscene 1935 version of 'Shave 'em Dry', which the folk blues guitarist Elijah Wald said was 'by far the most explicit blues song preserved at a commercial pre-war recording session'. The lyrics include, 'I got nipples on my titties big as the end of my thumb, I got something between my legs that'll make a dead man come.'

Dirty blues (sometimes known as bawdy blues) was a form of blues music that dealt with socially taboo and obscene subjects, often referring to sexual acts and drug use. Because of the graphic subject matter, most dirty blues was banned from radio and available only on jukeboxes. The style was most popular in the years before the Second World War, although it experienced a revival in the early fifties. The most particular aspect of dirty blues was the fact that it gave black women agency; most of the popular records were by female artists. While a pejorative interpretation is the reinforcement of the idea that black women were unnaturally keen on sex, it gave them a special kind of power. They owned innuendo. 'Need a Little Sugar in My Bowl' (which was subsequently recorded by Nina Simone) was written by Clarence Williams, who also wrote 'Organ Grinder Blues', which included the couplet, 'When you grind it slow, that's when I like it best.' Sugar had been a slang term for semen for over a hundred years, but its implication was often overlooked.

In 1955, the McGuire Sisters had a US No. 1 with 'Sugartime', and there were UK versions from Eve Boswell and Alma Cogan. It was written by Charlie Phillips and Odis Echols and the original lyric was written to amuse themselves while farming: 'Pussy in the morning, pussy in the evening, pussy at suppertime, be my little pussy, give me pussy all the time.' Bessie Smith died in 1937 after a car crash, possibly because she could not be admitted to a 'whites-only' hospital. She was buried in a pauper's grave and didn't have a headstone until Janis Joplin bought one in 1970. The inscription reads, 'The greatest blues singer in the world will never stop singing.' Also, the blues had a close connection with the LGBTQ community, and there was a long and decorated history of bawdy blues recordings aimed at a knowing gay audience.

Jackson, in the far more progressive seventies, wasn't trying to be licentious or prurient; she was acting like a documentarian, talking about things that weren't being discussed anywhere else. It was almost therapy. Consequently, it felt genuinely transgressive. Like Ma Rainey, she didn't sing in order to be scandalous, but to share what she saw, what she might have lived and what interested her about the needs of people she knew. She freely admitted to having affairs with married men ('Most of the time in my personal life, I was the other woman'), although was always keen to say she never wrote about her personal experiences. She was married herself, briefly (eight months) in 1971, to a bass player, who apparently thought they were going to be the next Ike and Tina Turner. 'He thought he was gonna tell me what to do with my life and I decided that was not gonna happen,' she said. 'Case closed.'

1975

These were conversations that women had with each other at the Laundromat, at home or in bars, she explained. You didn't hear these conversations on records. You especially didn't hear them on the radio.

At the start of the recording process, the idea was to try and capture Jackson's live act, but as soon as they got in the studio, the vibe was, 'Just do it'. None of it was planned. Her newfound popularity was such that she could start filling larger venues, but they were mainly full of women. Men did not especially want her records in the house. 'Because I spoke truth to women, I got a reputation for being rough on men,' she said. Some men found her uncouth, lowbrow and downright offensive – and women loved every millisecond of it.

Those live shows, especially during 1975, were incendiary. As the crowd whooped and hollered, encouraging her to go further in the exploration of her theme, her chosen subject, she would call for calm, and then almost incidentally say, 'This ain't for church folk!' It wasn't, and never would be. At the time, she could be guaranteed that there would be a number of 'walkers' at every show; one critic who saw her perform in the UK at the time said he noticed a smattering of people leaving each performance, presumably either bored or offended. As well as going into great details about the sex lives of the characters in her songs (sometimes unwisely using the gift of mime), she would instruct the audience in the correct way to swear. Comfort was not part of her repertoire. Emotion was. 'Her voice is amazing,' said one critic at the time. 'She sings like a man. An old, gold soul man. (Or, to avoid feminist backlash, perhaps I should say, like the R&B ladies of yesteryear – which amounts to the same thing).' Some reviewers (male reviewers) would start to be slightly patronising – one review from the British soul magazine *Blues & Soul* started in a typically blasé way: 'Groin-grinding, foul-mouthed, outrageously sensual, Millie Jackson isn't exactly the kind of girl you would take home to meet Mum – but she sure is fun' – and yet her influence would continue to grow. And it was all because of 1975, a year when the extremities of all kinds of music were creating an orthodoxy, a unity, of purpose. Excellence.

Saturday Night Special

11 October, Studio 8H, 30 Rockefeller Plaza, New York

SATURDAYS WERE INVENTED IN 1975, at least in the US. Before 11 October, late night television was very much the domain of Johnny Carson, where he ruled during the week, and where he was syndicated at the weekend (with The Best of Carson*). But when the chat show host decided he wanted to take more holiday, he asked for the compilation show to be used during the week instead, thus creating a scheduling vacuum on Saturday nights. And so, at the behest of media executive Barry Diller and various NBC bigwigs, Lorne Michaels – a man known for his deadpan humour, scriptwriting chops and intimidating stature – spent three weeks inventing* Saturday Night Live. *And in a heartbeat he made coast-to-coast stars out of Dan Aykroyd, George Carlin, John Belushi, Chevy Chase and Gilda Radner.*

Its pulling power was impressive. Guests on SNL *during the first series included Randy Newman, Raquel Welch, Carly Simon, Patti Smith, Kris Kristofferson, Janis Ian, Richard Pryor, Phoebe Snow, Gil Scott-Heron, and Peter Cook and Dudley Moore. The show was also nearly responsible for the return of the Beatles. Or at least two of them.*

On 24 April 1976, towards the end of the first series, John Lennon and Paul McCartney nearly took Lorne Michaels up on his offer to have the Beatles perform on Saturday Night Live. *Michaels had spoken directly into the camera about how the Beatles had affected his and his generation's lives. 'In my book, the Beatles are the best thing that ever happened to music. It goes even deeper than that – you're not just a musical group, you're a part of us. We grew up with you.' He then made his pitch. 'Now, we've heard and read a lot about personality and legal conflicts that might prevent you guys from reuniting. That's something which is none of my business. That's a personal problem. You guys will have to handle that. But it's also been said that no one has yet to come up with enough money to satisfy you. Well, if it's money you want, there's no problem here. The National Broadcasting Company has authorised me to offer you this cheque to be on our show. A certified check for three thousand dollars [about sixteen thousand dollars today].' The camera then zoomed in on the piece of paper in Michaels' hand. 'All you have to do is sing three Beatles songs,' he continued. '"She Loves You," yeah, yeah, yeah – that's a thousand dollars right there. You know the words. It'll be easy. Like I said, this is made out to 'the Beatles'. You divide it any way you want. If you want to give Ringo less, that's up to you. I'd rather not get involved.'*

A few blocks uptown, Lennon and McCartney were watching the show (along with twenty-two million other people) in Lennon's apartment. They were drinking beer and eating pizza, probably like a lot of people watching the show that night. As Lennon said in 1980, 'Paul was visiting us at our place in the Dakota. We were watching it and almost went down to the studio, just as a gag. We nearly got into a cab, but we were actually too tired. He and I were just sitting there watching the show, and we went, "Ha-ha, wouldn't it be funny if we went down?" But we didn't.'

McCartney confirmed this. 'John said, "We should go down, just you and me. There's only two of us so we'll take half the money." And for a second . . . But it would have been work, and we were having a night off, so we elected not to go. It was a nice idea – we nearly did it.'

A month later, on 22 May, Michaels returned with a new offer. 'We've heard from the Monkees, Freddie and the Dreamers, Herman's Hermits, Peter and Gordon, the Cowsills and Lulu,' he said. 'But still no word from the Beatles. I'm not discouraged, and neither is NBC. Because of the recent acclaim that Saturday Night *has received, I was able to convince NBC to sweeten the pot. John, Paul, George and Ringo – we are now prepared to up the original offer to $3,200.' He also offered free hotel* *accommodation at the Cross Town Motor Inn, which had a 'round-the-clock elevator service', water glasses than had been 'sanitised for your convenience', forty-eight-hour dry cleaning and free room-to-room calls.*

Six months later, George Harrison tried to claim the prize. Appearing on the show with Paul Simon, Harrison is shown haggling with Michaels about the amount. Saying he felt 'terrible' about the misunderstanding, Michaels said, 'I thought that you would understand, you know, that it was three thousand dollars for four people, that it would just be $750 for each of you. I mean, as far as I'm concerned, I mean, you could have the full three thousand. But the network . . .' Harrison responded by calling NBC 'chintzy', but, for an extra $250, was willing to say the show's traditional opening, 'Live from New York, it's Saturday Night!'

Saturday Night Live *was an instant hit, as popular with critics as it was with the public. So popular was it that it completely altered how people imagined their Saturday nights. 'Yes, it would be great to see the Hendersons, and they have nice wine, but do they watch* SNL?*' The show wasn't just a place for random political sketches, it was a satirical op-ed on American mores as well as a place for deconstructing popular culture and socio-political disparity. The impact of* SNL *– political, cultural, comedic – was as immediate as it was transformative. It didn't just surf the zeitgeist; it created its own ocean. New York City, the city that never slept, was suddenly pulsating with life and energy, on a television show that managed to be as exciting, as transgressive and as funny as the town itself. As a cultural phenomenon,* SNL *not only created an indelible mark on the world of comedy but quickly became a significant part of Manhattan's vibrant entertainment scene. And it very nearly got a couple of Beatles off the sofa.*

A Man Called Alias Torn Asunder

Blood on the Tracks is the ultimate, naked, full-frontal confessional. While John Lennon bared his own secrets to an almost unpalatable level – his *Plastic Ono Band* album was always a little overcooked – Dylan's only seventies' masterpiece was a kaleidoscopic emotional journey. 'A lot of people tell me they enjoy that album,' said Dylan. 'It's hard for me to relate to that . . . I mean, you know, people enjoying that type of pain.'

Blood on the Tracks by Bob Dylan

BLOOD ON THE TRACKS WAS allegory as avowal, therapy as entertainment. This remains his high-water mark, and for many it remains Dylan's very best record. For some at least it remains Dylan's last great record, the precursor to nearly fifty years of political conservatism, religious conversion, legendarily erratic stage performances and a release schedule where contemporary recordings were often completely overshadowed by compilations of often extraordinary unreleased material. Dylan has not always been his best editor, but on *Blood on the Tracks* he – perhaps inadvertently – managed to deliver what the world still recognises as the product of a genius. Here was an artist in torment, making the personal public, turning a flesh wound into art.

The year inspired an excessive number of records that went on to become an integral part of our classic rock canon. *Blood on the Tracks* was one of them; it was Bob Dylan's countercultural renaissance, beach-body ready after a smashing refit. It could also have been called *Divorce and the Art of the Impossible.*

When the record came out, and Dylan got wind of the feedback, he was confused. Simply couldn't understand it. Didn't people get it? Didn't they understand the torment that had led to his creative rebirth? But while the lyrics were coruscating and vicious and artfully delivered, the music soared, the evidence of a comprehensive return to glory, and Dylan's best album since *Blonde on Blonde*, nine years earlier. Some records are immortal, and this was one of those. Hilariously, he says the record wasn't autobiographical – going as far as to say it was inspired by the short stories of Chekhov – and yet it was the most personal album he'd ever recorded, full of first-person narratives and heartache, an album almost completely bereft of metaphor. Jakob Dylan, his son, said the album was the sound 'of my parents talking'. His pain, and the way in which he channelled that pain, produced the very best album of 1975, the very best Dylan album of the decade and possibly the best Dylan of all time.

Dylan began recording the album in New York City in September 1974. Before this, he had paid a visit to Michael Bloomfield, the electric guitar hero identified with Dylan's most rousing sixties' triumphs, and played him his new songs. But Bloomfield couldn't click with the new material, so the great reunion was not to be. Bloomfield later recalled the experience: 'They all began to sound the same to me; they were all in the same key, they were all long. It was one of the strangest experiences of my life. He was sort of pissed off that I didn't pick it up.'

Dylan slowly assembled a different group of musicians, and set to work. It was a difficult process, as he appeared to be making up the songs as he went, giving the musicians only vague instructions, and acting obtuse. 'Musicians dropped liked swatted bugs, writhing on the ground, waiting to die,' said Peter Brown, the engineer on the sessions. 'Studio musicians are tough; they're hired to do whatever it takes. When we would record with guys like Steely Dan, you might work on a basic track for twelve hours, searching for an impossible perfection, and you'd never say, "No," or show the slightest bit of attitude. But that was the game. This hurt. You could see it in the musicians' eyes, as they sat silently behind their instruments, forced not to play by the mercurial whim of the guy painting his masterpiece with fingerpaints.'

Nevertheless, eventually he finished it.

In December, shortly before Columbia was due to release the album, Dylan abruptly re-recorded much of the material in Minneapolis. He had played the test pressing to his brother, David Zimmerman, who persuaded Dylan the album would not sell because the overall sound was too stark. So, at his brother's behest, he agreed to re-record five of the album's songs, with backing musicians recruited by David. The new takes were done in two days at the end of December, as mournfulness and wistfulness gave way to a feisty, almost festive air (although with the album covers already printed, only the original studio band were credited). The sound was sweeter, but the pain was just as hard. He had changed some of the lyrics – the official version of 'If You See Her Say Hello' replaced its most heartbreaking line ('If you're making love to her/Kiss her for the kid' becoming 'If you get close to her/Kiss her once for me') – but the sheer resignation at work here still had the capacity to floor the listener. Dylan had turned the crisis of a deteriorating relationship into one of rock's most compelling dramas. During his separation from his wife Sara, Dylan had started writing songs, the kind of which would reconfirm his stature in a monumental way.

Blood on the Tracks is one of the most truthful dissections of love gone wrong, by turns recriminatory, bitter and heartbroken. The album was composed as Dylan's twelve-year marriage began to unravel, and songs like 'Tangled Up in Blue' and 'Shelter from the Storm' became templates for multidimensional, adult songs of love and loss. It is one of Dylan's peaks, the record where his genius and frail humanity finally meet. 'Tangled Up in Blue' would soon become most lay people's favourite Dylan song, although the ridiculously catchy music obscured the narratological high-wire act. Inspired by cubist painters, Dylan offered a seemingly simple story of a relationship, shifting in time and place and perspective like an old memory rescued from the past. *Blood on the Tracks* was the sound of real life catching up with him. For eight years, Dylan had been attempting to evade fame, focusing on his home life, while occasionally using domesticity as a platform to make records. Trying hard to be a good husband, music ceased to matter. For three years in the early seventies, he released nothing at all. At one time rock's untouchable king, he seemed washed up, bereft of ideas or ambition. With awful irony, it would take his marriage falling apart to rekindle his art. *Blood on the Tracks* was the record he pulled from the wreckage, having survived the initial crash.

'Having children changed my life and segregated me from just about everything that was going on,' he recalled in his book *Chronicles*. 'Outside my family, nothing held any real interest for me...I was fantasising about a nine-to-five existence, a house on a tree-lined block with a white picket fence...That would have been nice. That was my deepest dream.'

Demotivated by fame's assault on his everyday life, and resentful of his fans' delusional expectations, Dylan resolved to demolish his identity, transforming his image from rock and roll messiah to the happy hillbilly of *Nashville Skyline*. 'It's hard to live like this,' he said, recalling that time in the sixties when he was so feted that he was often ascribed ridiculous powers. 'The first thing that has to go is any form of artistic self-expression that's dear to you...Art is unimportant next to life...I had no hunger for it any more, anyway.'

Blood on the Tracks was Bob Dylan all grown up. Throughout the sixties he was so far ahead of the pack, so gifted a lyricist and jester than he could create smokescreens at will; no one could outsmart him, outrun him. And yet when his life became the victim of actuality, his talent followed suit. All of a sudden, allegory seemed fantastically old-fashioned; if you couldn't write about the maturity of anxiety then were you a real poet at all? Probably not.

Just before a February show in 1974 with the Band in Oakland, Dylan had met a twenty-four-year-old woman named Ellen Bernstein, who was running Columbia's A&R office in San Francisco. The two left a party thrown by promoter Bill Graham that night and went to her house, where they stayed up all night drinking and playing backgammon. Bernstein wasn't sure she'd hear from Dylan again, but soon after, he asked her to visit him at his house in Malibu. In the summer, he invited her to his farm in Minnesota, where he would spend his mornings writing lyrics in a red notebook. Bernstein said that Dylan would 'materialise around midday, come downstairs and eventually, during the day, share what he had written. It was in the notebook, but he would play it, and ask me what I thought, and it was always different, every time, he would just change it and change it and change it.' Many of these songs would never stop changing, as the emphasis on every interpretation was slightly different, sometimes radically, which is apparent from the demo tapes. 'It doesn't stop,' Dylan told Paul Zollo in 1991 of the differing versions of 'Idiot Wind'. 'It's something that could be a work continually in progress.'

Dylan had been lying about himself his entire life, but when real life interrupted his fantasy, he was forced to adapt. Luckily for him, his talent came with him. If he had written and recorded a song like 'Idiot Wind' ten years previously, for instance, it would have sat alongside the likes of 'Positively 4th Street' as a broadside against the societal swirl around him; in 1975 it felt as though he had almost physically removed the song from his own body and forced his band to somehow record it without him exploding with rage. At the time, it felt almost as important as 'Like a Rolling Stone'; with hindsight it might even be better than that.

'I thought I might have gone a little bit too far with "Idiot Wind",' said Dylan. 'I didn't really think I was giving away too much; I thought that it *seemed* so personal that people would think it was about so-and-so who was close to me. It wasn't . . . I didn't feel that one was too personal, but I felt it *seemed* it was too personal. Which might be the same thing, I don't know.' Sinéad O'Connor would later cite the song as a massive influence, saying that 'none of us would like to be the person he's talking to. That's why I love Bob Dylan . . . He can be real fucking nasty.'

Many critics couldn't believe the quality of the record, and so questioned themselves, thinking they may have made some mistake. *Rolling Stone*'s Jon Landau was one, although he did have this to say:

'The writing is the source of the record's power. It's been a long time since Dylan has composed a melody line as perfectly suited to his voice as "Tangled Up in Blue", and though the lyrics are both confessional and narrative, Dylan makes it all sound like direct address. There are times when he sounds closer, more intimate and more real than anyone else.'

Which was undoubtedly because he was now writing about his own problems rather than someone else's. What was once personal was now public, and he was making it so. Which was a reflection of society at large as much as it was a reflection of himself. If the so-called 'free love' explosion of the sixties had changed teenage social contracts, by the mid-seventies it had entered the mainstream. Lifestyle expectations were suddenly off the charts. In 1973, Alex Comfort's *The Joy of Sex* had become a suburban bestseller, while Hollywood was finally reflecting the realities of marriage. In literature, John Updike's 1968 novel *Couples*, which depicted the lives of a promiscuous circle of ten couples in the small Massachusetts town of Tarbox, was an examination of the 'post-pill paradise' sweeping America. At the time this was still deemed transgressive; *Time* magazine had reserved a cover story for Updike and the novel before knowing what it was about; after actually reading it they were embarrassed as they discovered, according to Updike, that 'the higher up it went in the *Time* hierarchy, the less they liked it'. The sexual revolution that had supercharged the sixties resulted in the marital tumult of the Swinging Seventies. And now people were even making pop music about divorce.

In 1969, legislation on both sides of the Atlantic had had a huge effect on divorce rates. In the UK, the Divorce Reform Act marked an important shift by adding further grounds for divorce – allowing divorce on the basis of two years' separation with the other party's consent or five years' without. It also removed the concept of 'matrimonial offences', such as adultery, desertion or cruelty toward one party, and hence the idea of divorce as a remedy for the innocent against the guilty. These liberalisations, combined with changing attitudes and expectations of marriage, and the greater economic independence of women, all contributed to a rise in the number of divorces from 50,000 per year in 1971 to 120,000 four years later. In the US, Governor Ronald Reagan of California sought to eliminate the strife and deception often associated with the legal regime of 'fault-based' divorce, signing the nation's first no-fault divorce bill. The new law eliminated the need for couples to fabricate spousal wrongdoing in pursuit of a divorce. Because this was California, this new development was viewed as a lifestyle opportunity. Dr Clinton E. Phillips, a sociologist, said at the time, 'In California we have a highly mobile, anonymous, transient, striving, thrill-seeking, experience-seeking population. All of this has an effect. Divorce is much higher in large cities than in small towns, and divorce goes along with affluence – people can afford it. There is also the influence here of the climate; it is possible to get out and socialise more here than back east. Another factor is that people live longer now – they have more time to get tired of each other.'

In the years that followed, virtually every state in the union followed California's lead and enacted a no-fault divorce law of its own. By the end of 1975, the number of divorces in America in one year would pass the one million mark for the first time, which was more than double the 479,000 divorces recorded in 1965. In 1977, Avery Corman would write the best-selling *Kramer Versus Kramer*, about an estranged couple locked in a custody battle over their young son. Two years later it was turned into a movie starring Dustin Hoffman and Meryl Streep. It would go on to win five Academy Awards, including Best Picture. For some reason nothing from *Blood on the Tracks* was on the soundtrack.

Dylan wasn't alone, of course. A year earlier, Willie Nelson had written an entire album about divorce, *Phases and Stages*, a concept record anchored by the repetition of its title track and the split narrative – side one told the story from the woman's perspective, side two told the story from the man's. This would be followed by Marvin Gaye's *Here, My Dear*, Fleetwood Mac's *Rumours* and eventually Bruce Springsteen's harrowing *Tunnel of Love*. Country artists had been using spousal separation as a legitimate seam for decades, while Frank Sinatra had made the torch song his own – after divorcing his first wife, he married Ava Gardner, kicking off a turbulent marriage that lasted until 1957; in the middle of that, he had crafted *In The Wee Small Hours*, an album that meditates on loss and loneliness and moving on from broken relationships. But it took Bob Dylan to really turn divorce into an art form.

Dylan's 'comeback' album never changes, never fades and rarely disappoints anyone coming to it for the first time. It continues to be a source of fascination for Dylan fans and musical scholars, an unyielding monument to the power of heartache and pain. '*Blood on the Tracks* is the best album of 1975, as there are so many great tracks on it,' Courtney Love told me. 'There is so much good work on it. "Shelter from the Storm". I'm dead, I'm fucking dead, that song kills me every time. "Tangled Up in Blue", the end. Every song is a work of genius.'

'This is my favourite album ever,' said Quentin Tarantino. 'I spent the end of my teenage years and my early twenties listening to old music – rockabilly music, stuff like that. Then I discovered folk music when I was twenty-five, and that led me to Dylan. He totally blew me away with this. It's like the great album from the second period, y'know? He did that first run of albums in the sixties, then he started doing his less troublesome albums – and out of that comes *Blood on the Tracks*. It's his masterpiece.'

The album made a huge impression on me, largely because it was the first Bob Dylan record I had any real relationship with me. I was too young for the sixties, and as Bowie had said in 'All the Young Dudes', I'd never got it off on that revolution stuff. For Bobheads, *Tracks* was a return to form, a way to negate everything from *John Wesley Harding* to *Planet Waves*; to people from my generation, who were ten to fifteen years younger, *Tracks* was an introduction rather than a restoration.

It was David Bailey's favourite Dylan album too. Dylan always wore his persona like a shroud, rarely taking it off. Bailey remembered him turning up to be photographed at his Bloomsbury studio in London and it was almost as though he was a ghost. 'He was there all day, pretty much, although he hardly spoke at all,' Bailey told me. 'He was grumpy, even grumpier than me, as I think he had woman problems at the time. He was grumpy until about four o'clock in the afternoon when he started emerging from his haze. One of my children was with me and if you're a kid and you see someone dressed in a tasseled leather jacket and eyeliner, you're going to stare. Dylan kind of warmed to that. He started to stare back, and immediately made a connection. He was one of these people with a completely "Couldn't give a fuck" attitude. As well as being a genius.'

If art is a means of making sense of experience, then *Blood on the Tracks* was art.

Nova and Biba

14 October, King's Reach Tower, London SE1

THE SIXTIES OFFICIALLY ENDED IN October, as towards the middle of the month, Nova *magazine closed its doors for the very last time, after ten years. Biba had already shut its doors a month earlier.*

In the summer of 1963, fashion illustrator Barbara Hulanicki established a mail-order company, which she named Biba, selling affordable fashion appealing to a new generation of young women. Their postal boutique had its first significant success in May 1964, when it started selling a pink gingham dress with a hole cut out of the back of the neck with a matching triangular kerchief, to readers of the Daily Mirror. *The dress had celebrity appeal, as a similar garment had been worn by Brigitte Bardot and more than four thousand orders were received on the first day. Ultimately, some seventeen thousand outfits were sold. The brand's first store, in Abingdon Road in Kensington, opened the following year, and at the time there was nothing like it in London. 'It isn't just selling dresses, it's a whole way of life,' said Hulanicki. Big Biba, the final iteration of the brand, opened in Kensington High Street in September 1973, a simply extraordinary department store which would soon have its own restaurant on the roof, complete with pink flamingos. When the Roof Garden opened in May 1974, Biba threw one of the first great parties of the decade; writing about it in the* Daily Express, *Sandy Fawkes said it 'was like walking into a film set of rather jolly moral depravity', much like Biba itself.*

When Biba launched it was an independent company; in 1969 a new company was formed and the majority of the shares sold to Dorothy Perkins, which in turn became a wholly owned subsidiary of the property firm British Land in August 1973, just a month before the store opened. And what a place Biba was: at the time the most extravagant shop in London. It wasn't just huge, it was cool (or, as people would have said at the time, 'trendy'; no one said 'cool'). Like a Roxy Music album cover, in fact. Also, there was nothing Biba didn't appear to sell; from cosmetics and food to books and lots of things most people didn't know the names of. Some thought it was just a clothes shop, whereas it was a lifestyle squeezed into a building. The food stall was amazing enough by itself, selling rather marvellous black-and-gold tins of consommé, shark's fin soup, bird's nest soup, real turtle soup, lobster soup, vichyssoise and baked beans.

The timing of the new opening couldn't have been worse. The oil crisis, the rise in world commodity prices, the miners' strike and the three-day week combined to produce a property slump that reduced British Land's shares to just ten per cent of their value within just a few months. Big Biba was used as a way to drive profits, when what it needed was time to bed in. The inevitable happened in 1974, when relations between Hulanicki and her husband, Fitz, and British Land collapsed, and the building suddenly had a target on its back. It staggered on for a while, but the company soon closed, and with it the last vestiges of the Swinging Sixties. It appeared the revolution would not be stylised.

Nova *was first published in March 1965, a British glossy that was described by* The Times *as 'a politically radical, beautifully designed, intellectual women's magazine'. It covered such once-taboo subjects as abortion, cancer, the birth-control pill, race, homosexuality, divorce and royal affairs. It featured stylish and provocative cover images and was immediately the most fashionable magazine in the world. Nova ran five-thousand-word articles by Christopher Booker, Susan Sontag and Irma Kurtz with photography by Helmut Newton and Don McCullin, and fashion by Caroline Baker and Molly Parkin. Running with the tagline 'The new kind of magazine for a new kind of woman', Nova set itself apart from its contemporaries by creating a magazine for an audience that was not only interested in fashion, but one that was also politically, socially and sexually aware.*

Looking back now, it almost seems as though everything happened at once. In a decade dominated by youth, London had burst into bloom. It was swinging, and it was the scene. The Union Jack suddenly became as ubiquitous as the black cab or the red Routemaster, and all became icons of the city. Carnaby Street's turnover was more than five million pounds in 1966 alone. Quite simply, London was where it was at. Fuelled by growing prosperity, social mobility, post-war optimism and wave after wave of youthful enterprise, the city captured the imagination of the world's media. Here was the centre of the sexual revolution – the pill had been introduced in 1961 – the musical revolution, the sartorial revolution. London was a veritable cauldron of benign revolt. And at its heart was Nova, *the coolest magazine in the world.*

Nova *was known for its unflinching confidence and its ability to tackle controversial and important societal issues with its bold and dynamic editorial approach. It represented a new type of features magazine which sought to address readers by their interests and attitudes rather than their age, class or socio-economic groupings. Its publisher, IPC, hoped to capture an affluent and socially mobile female consumer, women with a sense of enquiry, intelligence, humour and ambition. It looked extraordinary, too, aiming to shock, to provoke, and (important this) make you laugh. It luxuriated in empty space, bold print and experimental photography. It was, in some sense, a smorgasbord of novelty for the chic, progressive middle classes. Few magazines exerted so much influence at the time, or indeed now. When it launched, it flew, and during the first two years of publication it was selling upwards of 150,000 copies a month. Ten years later, with a lack of managerial support, and a shift in the zeitgeist (everyone else had caught up), it was barely selling half this amount. Like Biba,* Nova *offered a capsule lifestyle, a pick 'n' mix alternative to anything else to be found on the high street. But by the end of the year, they, like the sixties, were finally dead.*

Superfly Meets
Margaret Bourke-White

Like many albums released in the seventies, this was known as much
for its cover as what was inside. It was based on a 1937 monochrome
photograph by Margaret Bourke-White, *At the Time of the Louisville
Flood*, which was originally published in *Life* magazine in 1937. It was
the perfect wrapping. If you squinted, the mid-seventies was a period
of destitution, paranoia and urban neglect, with race at the centre of
it all. The message of Mayfield's album was clear. Forty-five years on,
and what's changed?

There's No Place Like America Today by Curtis Mayfield

A LINE OF FIGURES QUEUES across the frame on the big city sidewalk, all of them in hats and overcoats, some carrying empty bags
and buckets, every one of them black. A few look at the camera, but
none with enthusiasm. They are flood victims lining up to get food
and clothing from a Red Cross relief station. They are standing in front
of a gigantic poster produced by the National Association of Manufac-
turers, of a cheerful, white, middle-class family in a new car, under a
banner that proclaims, 'World's highest standard of living', accompanied by the strap-line, 'There's no way like the American Way'. Taken
by the great photo-journalist Margaret Bourke-White, the picture –
titled *At the Time of the Louisville Flood* – appeared in the 15 February
1937 issue of *Life* magazine, in the days when an image in *Life* was seen
by most people in America.

When the Ohio River flooded Louisville, Kentucky, Bourke-White was sent to the area to get as many pictures as possible. Documenting what was one of the largest natural disasters in the history of the United States, her image offered a commentary on perceived racial and economic inequities. Ranking alongside the likes of Arthur Rothstein and the work of the FSA photographers (who documented the devastation of the dust bowl earlier in the decade), *At the Time of the Louisville Flood* had achieved iconic status long before it was adapted for the cover of Curtis Mayfield's 1975 album. Mayfield chose it for very obvious reasons; it wasn't subtle, but it worked. Just like the best work of any protest singer.

The world Mayfield's record portrayed was very cold indeed and the comfort it offered was as spare as its measured, melancholy funk. A 'message' record, it was compared favourably to Sly Stone's *There's a Riot Goin' On*, the O'Jays' *Ship Ahoy*, Stevie Wonder's *Fulfillingness' First Finale* and Marvin Gaye's *What's Going On*. Obviously, it wasn't easy being compared to the likes of Sly, Stevie and Marvin, but the critics had a soft spot for Curtis. He was perceived to be a benign presence, one of the good guys, so everyone cut him some slack. He was dubbed the 'gentle genius'. His talent might not have been as obvious as the other guys', but he was in the club. After all, he'd written 'People Get Ready' for the Impressions, and 'Superfly' and 'Move on Up' for himself, so he couldn't have been that bad. No lesser an authority than Martin Luther King Jr named 'People Get Ready' the unofficial anthem of the civil rights movement, even though the lyrics didn't address the state of race relations as explicitly as some other songs. 'Move On Up' was Mayfield's American optimism for a new decade – tougher, faster, funkier. The chugging guitar and flared horns nodded to James Brown and the insistent congas kept an open ear to the widening musical dialogue between Africa and the rest of the world. The song was framed as a conversation with a child, Mayfield the avuncular sage who knows just what to say when times seem overwhelmingly tough.

The album was written in Atlanta in early 1975, during two weeks of terrible depression. 'It was a long way from *Super Fly*,' said his son and biographer Todd, 'a hard look at some of the things that sour our life experience.' The record was as much about Mayfield's mental health as the parlous state of America. It was called a funereal, seven-song state-of-the-union blues, focusing on murder, depression, fear, love, paranoia, prayer and poverty. *MOJO*'s Andrew Male said: 'Suffused in vulnerability and defeat, it might also be his most confident album, utterly certain of its lyrical message and its sedate, slow-burn power.'

The hypnotic 'Billy Jack' (about a convict, on the day of his release, finding out about the tragic shooting of an old friend), was the opening song, a spellbinding, lazy, funk groove with a deliberately sparse arrangement. Like Mayfield's interpretation of 1975 itself, it was desperate: shimmering wah-wah guitars, thick bass triplets, low-fi brass and Curtis's broken falsetto. It was downbeat for a reason. As a young man, Mayfield had witnessed the exploitation of many an (often illiterate) blues musician in Chicago as he was coming up in the world. Consequently, he was one of the first black American musical figures to call out the implicit racism in the music industry. He was called 'the thinking man's soul man', a musician's shop steward. Always wise counsel. Mayfield's music focused on socio-political matters from the point of view of a storyteller and this record was his way of telling anyone who would listen that the country – his country – was still very much a developing nation. It touched on gun violence, economic woes and the struggles of life in the city, but the litany of problems was balanced by resilience and beauty, particularly on the singles 'So in Love' and 'Love to the People'.

Mayfield's reputation had been built on the protest songs with which he documented the sixties civil rights movement – 'Keep on Pushing', 'People Get Ready', 'We're a Winner' etc. – and here he was in 1975, doing it once more, only this time his optimism had deserted him. Ultimately, this was a suffocating portrait of America, and one that was unrelenting in its fatalism. Unlike his previous work, there was no honeyed veneer to sweeten the taste.

A few years later, Paul Weller (who would later successfully cover 'Move On Up') interviewed Mayfield at Ronnie Scott's, London. 'I like to go in depth as to where I know without a doubt that those who receive me understand me,' he told Weller. 'I know they breathe; I know they cry; I know they're hurt; I know they love; I know they hate. They have all these different feelings. When you speak in terms of depth rather than ride along the shallow surfaces, they can only give you one true reaction as to what you're talking about.'

'He was a genius,' said Weller. 'A lot of the things he was talking about back then are still relevant now. Racism, inequality, ecology, corporate takeover. I think of him as a prophet, but he's a beautiful, romantic writer as well. He covers the whole spectrum. Those solo records in the early seventies, it's just one classic after the other. *Curtis, Super Fly, There's No Place Like America Today.* It was a golden run, and a golden time.'

It wasn't a golden time in South Central LA in 1975, which felt as far away from Beverly Hills as the South Bronx. If Malibu and Bel Air were rich, South Central was decidedly poor. And the palms in Watts were very different from the ones in Pacific Palisades. South Central Los Angeles was often characterised as an African American community beset by poverty and economic neglect and by the seventies the area was in crisis. The area embodied a complex history that captured the dynamics of spatial inequality. African Americans had begun moving to LA in large numbers in the early twentieth century, and for the next forty years their numbers doubled every decade; by 1940 they represented slightly more than four per cent of the total population of the city. But the rapid decline of the area's manufacturing base (as companies such as Goodyear, Firestone and General Motors moved out) resulted in a loss of the jobs that had allowed skilled union workers to enjoy a middle-class lifestyle. Downtown Los Angeles's service sector, which had long been dominated by unionised African Americans earning relatively fair wages, replaced most black workers with newly arrived Mexican and Central American immigrants.

Widespread unemployment, poverty and street crime contributed to the rise of street gangs such as the Crips and the Bloods. Formed in 1969, the Crips were named after gang members began carrying around ornate canes to display their pimp status. People in the neighbourhood started calling them cripples or 'crips' for short. The Bloods (Black Liberation Organisation Of Defence) were formed four years later, an amalgamation of various smaller gangs, who had become sick of Crip dominance.

Following the 1965 Watts riots – six days of civil unrest that stemmed from a drunk-driving arrest – South Central became a hotbed of the civil rights movement. The reform of policies shaping segregation, immigration and urban investments in the sixties was profoundly changed the American urban landscape. Organisations like the Studio Watts Workshop, the Watts Writers Workshop, the Watts Towers Arts Centre and the Black Panthers utilised both the arts and direct action as legitimate methods of raising awareness and attempting to create social change.

By 1975, the atmosphere in South Central Los Angeles was nothing if not febrile, mirroring the mood in Washington, New York, Chicago, Detroit, Baltimore and San Francisco. Anger and tension didn't only inform the work of Curtis Mayfield, but that of everyone from Gil Scott-Heron to Melvin Van Peebles, from Funkadelic to Swamp Dogg, from Lou Rawls to the Bar-Keys, from James Brown to Isaac Hayes, from the Staple Singers to Earth, Wind and Fire. In their own way, they were all protest singers. In this world, segregation wasn't simply political, geographical, economic or sociological, it was existential, particularly in the music industry. Curtis Mayfield wanted to share messages of uplift that were more insistent and urgent than usual, and he needed more than a metaphor to get his messages across. He needed agitprop. Though he had mastered the three-minute single, like his peers he was now enjoying using the album to express his artistic vision. Commenting on record industry conventions in the sixties, he noted in 1975 that 'usually you put all your B-sides on an album, along with the single records you had put out. Very few people ever thought of recording an album as a complete concept, a story to tell from the first cut to the last to a point that even if you just read the titles of the songs, they would just about make up their own paragraph.' *There's No Place Like America Today* was one hell of a paragraph.

'Although it's never openly stated, *No Place Like America* portrays the comedown from the sixties,' wrote David Bennun, rather brilliantly, in *The Quietus*. 'For affluent white kids, that decade may have meant peace and love (i.e. plenty of drugs and unlimited opportunities for boys to sleep with girls too frightened of appearing repressed to complain about bad and/or unwanted sex), and the chance to play at being revolutionaries. For blacks, it meant political and social gains on a previously unimaginable scale; gains that by 1975 seemed to be reversed, in practice, at every turn. True, some of the promises of the sixties were fulfilled; America has developed a large black middle class, a process which was well under away by the mid-seventies. But it was those left behind who were the subject of Mayfield's album – the seventies' equivalents of the hard-worn folk on the cover.'

Mayfield, perhaps unlike most of his peers, appeared short of ego, at least the kind of ego that supplemented his cause with his own self-esteem. His message wasn't wrapped around his image, and his anger and disappointment weren't calibrated by intensity or amplified by messiah-like tendencies. At the time, it wasn't especially dignified for protest singers to occupy the foreground, as what they were selling was the message, not the messenger. Well, that was certainly the case with Curtis. Soon, the whole nature of African American anger would manifest itself in the cultural broadside of hip-hop, changing culture for ever: the idea of self would eventually envelop and minimise the root cause of the disappointment. While hip-hop would start as party communication and celebration, it would quickly morph into the strident and the violent. The third stage would acknowledge that we had moved into the secular age of identity, in which a person's selfhood was celebrated in the way more religious eras obsessed over the idea of the soul.

This felt like the end of something, as soon, protest songs would take on a wildly different form. If, in 1975, protest songs borne of the black experience were written and administered by the sagacious and the revered, they would soon be the domain of the young, the boisterous and the professionally aggrieved. In truth, hip-hop had already begun. It had started on 11 August 1973, when some teenagers threw a back-to-school party in the rec room of their apartment building in the Bronx. One of them was Cindy Campbell, a high-school student who, in order to be able to afford clothes for the new school year, decided to throw a party. She asked her eighteen-year-old brother Clive to DJ for her; his street name was 'Kool Herc'. At the time, like other parts of the city, including the Lower East Side, Bedford-Stuyvesant and Harlem, the Bronx looked like many bombed European cities after the Second World War. New York was almost bankrupt and, as there was really nothing much for kids to do, they became resourceful. As was Cindy.

Admission that night was 25 cents for girls and 50 cents for boys. In thirty-two-degree heat, Herc dragged his sound system, consisting of two turntables, a mixer, an amplifier and a set of speakers, into the white-walled community room of the tenement building at 1520 Sedgwick Avenue in the Morris Heights area. The smell of marijuana was in the air. The bass pumped and ripped through the sweating crowd as they danced and partied in the darkened room. Herc noticed that some of the dancers loved the parts of the records where the vocals dropped out, leaving just the percussive breakdowns or the breaks. He started playing the break from one record, then playing the break from another on his other turntable. He'd then queue up another break on his first turntable. He also soon figured out how to use his two turntables simultaneously to loop a single break with two copies of the same record, letting the dancers go crazy. He would call this technique the 'merry-go-round'. That night, Herc's buddy Coke La Rock grabbed the microphone, greeting Cindy's guests by name. This would later earn him a reputation as the first MC. That night, several hundred people turned up to dance, drink soda and sing along at what would become the very first hip-hop block party. There would be a lot more parties, and Herc never revealed which records he was using, to make sure people would keep coming to hear his exclusive sound.

In truth, he was no stranger to sound systems. Growing up in the Trenchtown neighbourhood of Kingston, Jamaica, young Clive had been immersed in U-Roy, the Skatalites and other local acts. After he arrived in New York in 1967, he danced at clubs, deejayed and did some graffiti art: the name 'Kool Herc' was a combination of the then-ubiquitous commercials for Kool cigarettes and his track-and-field skills (he was nicknamed 'Hercules'). His own DJ skills started at home. His father, a mechanic, already had a sound system; Herc himself would sometimes pull speakers out of abandoned cars. 'My father brought a PA system and didn't know how to hook it up,' he said. 'I was messing around with the music, and I started out by buying a few records to play at my house.'

That night he remembers playing 'Apache' by the Incredible Bongo Band, 'The Mexican' by Babe Ruth, 'It's Just Begun' by the Jimmy Castor Bunch, 'Melting Pot' by Booker T. & the M.G.'s, 'Give it Up or Turnit a Loose' by James Brown – and 'Move on Up' by Curtis Mayfield.

Street Life

I STILL HAVE THE COMPLETE set in my library, a pile of beautifully designed magazines that attempted to define the world as we knew and understood it in late 1975. Like many titles at the time, Street Life *was a thing of great beauty, launched with some fanfare in the UK in November, ostensibly as a British equivalent to* Rolling Stone *and designed to appeal to consumers of the arts as well as music journalism (basically the* NME*), Sunday supplements and* Time Out. *It was a full colour, bi-weekly tabloid deliberately aimed at the cultural connoisseur, the intellectual butterfly, aping the cod-newspaper graphics as a way to elevate the subject matter. It cost 25p, which was not cheap (the* NME *at the time was 12p). The magazine also had a great logo, an oval brick wall with 'Street Life' in a stylised spray-can script. It was distributed by Condé Nast, and funded, so many thought, by Chris Blackwell, which was borne out by the magazine's extensive reggae coverage. It was founded by Billy Walker and Andrew Sheehan, who had left the music weekly* Sounds *in the hope of producing something more artful.*

Like Rolling Stone, *it had a formal format, designed to look like a broadsheet newspaper. The covers were beautiful, and the publication covered a mix of music and culture, including an image of Pete Townshend in his white boiler suit by satirists Luck and Flaw, a Bob Lawrie drugs issue, a sci-fi special starring the Mekon, espionage ('it's a conspiracy theory!'), crime, Bryan Ferry the art collector, Jimi Hendrix the legacy, Stan Bowles the footballer, and the kind of front pages that assumed a greater level of engagement than usual. It was a proto-lifestyle magazine, a grown-up magazine aimed at the grown-ups who were consuming grown-up pop culture, principally music. It was mature, like its audience, like their tastes.*

The first issue had features on the disappearance of the Labour MP John Stonehouse (who unsuccessfully attempted to fake his own death in 1974), Lindsay Anderson's new play, pub rock stalwarts Kokomo, the Who, British boxing, a lukewarm piece on Fawlty Towers, *and a bunch of bad-tempered album reviews. Its mistake was forgetting to include the fun stuff, because while it was erudite, cool, and full of great writing, when you were flicking through it in the newsagents there were no titbits to make you stop. I bought every issue because I was obsessed with magazines, but most people at the time read publications because there was something (usually just one 'something') they were interested in. A magazine was like a sweet counter; it didn't matter how well-designed it was, what the rest of the sweets looked like and how much they were (no sweets were very expensive, in the same way that magazines were always great value for money); you were probably interested in one sweet and one sweet only.*

What is particularly interesting now is the coverage given to those acts who would soon fade from view, but whose profile at the time was disproportionately big. There were so many of them: the Sadista Sisters, Clancy, Be-Bop Deluxe, Moon, Mr Big, Outlaws, Splinter, Back Street Crawler, String Driven Thing, Murray Head, Magma, Frankie Miller, Steve Goodman, Boxer, John Stevens' Away, John Miles ('the voice of 1976'), Charlie, the Jess Roden Band, Mallard, Widowmaker, Rory Gallagher, Patrick Moraz, Michael Pinder, Druid, Sailor, Meal Ticket, Ice, Sassafras, Shanghai, Supercharge, Strapps, Streetwalkers, Pure Prairie League, Terry Reid, Michael Chapman . . . the bargain-bin makeweights that gave the mid-seventies such a bad name.

Groups like these would soon be obsolete, kicked into touch by the arrival of a bunch of smartarse guttersnipes managed by Malcolm McLaren. Street Life *actually interviewed McLaren in early 1976, in his shop in the 'sleazy' part of the Kings Road (Sex, formerly Let It Rock and*

Too Fast to Live, Too Young to Die), and although it mentioned his partner 'Viv' Westwood, it managed not to mention the Sex Pistols at all. McLaren explained that they had become bored with the fifties' nostalgia boom, and so closed the shop in order to do something else. He said they became hooked on the idea of opening a gymnasium where you could buy rubber suits and clothes for the body that you could sweat in. He also mentioned the suits, leather jackets, the 'political and sexual' torn T-shirts, the cock rings, masks and vinyl tops. You could tell from the interview that this was McLaren beginning to test the waters in terms of what he thought he might be able to get away with in interviews. As his notoriety grew, he would start to be more ambitious, and more outlandish.

'I've always been involved with cults – the subterranean influence on people – that's what fashion is predominantly about,' he said. 'The fashion market at the moment has separated the kids into all different factors. They can be fashionable either by going to the Portobello or a chain like Take Six or something chic like Sterling Cooper, because all commercialism feeds on diversification. Kids have a hankering to be part of a movement like the Teddy Boys of the fifties and the mods of the sixties. They want to be the same, to associate with a movement that's hard and tough and in the open like the clothes we're selling here.' The headline was WOULD YOU BUY A RUBBER T-SHIRT FROM THIS MAN? *It was more than a rhetorical question.*

Street Life *would have a short life, and would close the following summer, seventeen issues after launch, unable to woo readers from the* NME, *which was already successfully covering this particular waterfront (thank you very much). Ultimately it was too dry for British tastes, and too niche. Too diverse, too smart, and obviously ahead of its time,* Street Life *would soon be forgotten. It would be resurrected in a completely different form with* The Face, *five years down the road.*

A Velvet Dildo Pierces
the Leisure Society

If anyone anticipated Tom Wolfe's withering 'me decade' putdowns of the self-obsessed, designer renegades of 1975, it was Donald Fagen and Walter Becker's Steely Dan. The critics who initially lambasted them for making music for lotus eaters didn't register the innate sarcasm in their lyrics. Assuming they were assembled simply to please FM radio, consumed by a generation who had long since abandoned insurrection in favour of designer clothes and personal development, they were perceived as just another LA band. Couldn't have been more wrong.

Katy Lied by Steely Dan

SOCIAL ANTHROPOLOGY ENJOYED A GROWTH spurt in the seventies, as instantly recognisable tribes started to appear in the most unlikely places. While some were simply the natural development of those who had bought into the counterculture in the mid- to late-sixties, and who had decided to stay that way, others were the manifestation of a new consumer culture.

Those curious teenagers of the sixties were now financially empowered twentysomethings who wanted their cultural choices manifested by their lifestyle, not that anyone called it 'lifestyle' back then. The baby boomers who had driven the activism of the sixties were starting to settle down. As they did, they looked inward instead of outward. Fewer people protested in the streets, and many more visited therapists or sought to improve their spiritual lives as well as

287

their material lives. This quest for individual perfection led to a higher divorce rate, as people found it more acceptable to leave a marriage if they found it unfulfilling. The war in Vietnam, Watergate and the FBI and CIA scandals, had left the electorate so shell-shocked and disillusioned that they were now going to believe in themselves rather than other people.

Conspicuous consumption became its own kind of activism, in the form of status symbols (cars, fashion, furniture) and fads (aerobics, jogging, mood rings). If you were a 'dude', you probably had a deep tan, a Hawaiian shirt and a peach-coloured sweater – a dynamite colour combination, all very casual and spontaneous. If you were a 'chick', you had stopped looking like Carole King and were now trying to look like one of the Pointer Sisters. It was also the beginning of a bespoke world, exemplified by copywriter Shirley Polykoff, who worked for the ad agency Foote, Cone & Belding. One of her accounts was Clairol hair dye, for whom she came up with a slogan which immediately quadrupled the brand's product sales: 'If I've only one life, let me live it as a blonde!' In a single catchphrase she summed up the decade.

As the sixties sleepwalked into the seventies, you could be forgiven for thinking that many people figured that the best way to empower their creativity was through an infusion of good taste. You could almost imagine the modern consumer warming the bowl of their wineglass in the palm of their hand, listening to a playback of their newly purchased albums in the music room in their split-level lateral dreamhome and thinking, Hmmm, sophisticated . . .

In 1975, Cyra McFadden would start planning *The Serial*, a book satirising the self-obsessed denizens of California's Marin County, where everyone talked in the psychobabble of faddish self-help manuals while keeping up appearances via a mellow diet of TM, lentil loaves, Zen jogging and vintage tennis shirts. In this new 'Me' decade, as Tom Wolfe had noted, the alchemic dream was remodelling your personality along with your clothes, your beliefs and what they were now encouraged to call 'lifestyle'. At a Marin County wedding, one of McFadden's protagonists spots a friend wearing 'Marie Antoinette milkmaid but with her usual infallible chic. She had embellished it with her trademark jewellery: an authentic squash-blossom necklace, three free-form rings bought from a creative artisan at the Mill Valley Art Festival on the right hand, and her old high school charm bracelet updated with the addition of a tiny coke spoon.'

Music became commodified, too, as the strident posturing of the late sixties morphed into a revolution defined by production values and virtuosity. In the sixties you could get away with drumming like Ringo Starr; by 1975, if you didn't play like Steve Gadd (who as well as playing with Steely Dan appeared on Paul Simon's '50 Ways to Leave Your Lover', Van McCoy's 'The Hustle' and Rickie Lee Jones's 'Chuck E.'s in Love') you were considered to be well below the salt. As they strove for the kind of sonic perfection that had been made more attainable by the rapid transformation of the seventies' recording studio, becoming impossibly hard taskmasters in the process, so Steely Dan's Donald Fagen and Walter Becker also pursued an image of disaffected cool, openly disparaging the industry that had embraced them. Ungrateful to the core, they nevertheless produced some of the most sophisticated music of the era.

Exhibit A: *Katy Lied*, as gentrified and as anal a record as you'll ever hope to hear. Steely Dan's 1975 masterpiece is an homage to West Coast passive-aggressive studio cool, even though they were completely disdainful of the palm tree and flared-denim world of Los Angeles. The band's nihilism is plain for all to hear, disguised as FM-friendly soft rock. Donald Fagen's lyrics are dispassionate, the architecture of their songs often labyrinthine, the guitar solos ridiculously sarcastic. Yet they made some of the most sophisticated, most polished, most burnished music ever heard.

Fagen and Walter Becker were scathing about the hard-rock world – finding groups like Led Zeppelin and Bad Company preposterous – and were far more interested in the construction of old jazz records. For them, the only correct response to the entire culture of 'rock' was to be dismissive. They were occasionally, and unfairly, compared to the soporific jazz-rock that seeped across US radio in the seventies, as their obsession with technical proficiency was mistaken for musical indolence.

And *Katy Lied* was really where it all started. The band had made their opening remarks in 1972's extraordinarily accomplished *Can't Buy A Thrill* (where they sounded like just another bunch of talented if cynical West Coast musos), and then followed it with 1973's *Countdown To Ecstasy* (more cynicism, this time with a darker edge) and 1974's breakthrough album *Pretzel Logic* (where they showed off their jazz leanings on a capsule collection of great pop songs). But with *Katy Lied* they finally married their elegant misanthropy with studied and refined musical seamanship, fusing their mordant wit and burnished jazz-pop. With this album they finally conflated the various ways they had been trying to define themselves, on the one hand showing their prowess as smartass unbelievers, and on the other using expertise and virtuosity almost as weapons.

Steely Dan were the musical personification of the iron fist in a velvet glove, and if their lyrics didn't kill you then the jazz licks would. All of the band's albums are finely calibrated dissertations on the Me Generation's human condition, and *Katy Lied* is where they pivoted.

Donald Fagen met Walter Becker while studying at Bard College, a private liberal arts college in Annandale-On-Hudson in New York State, in the mid-sixties. He was nineteen, Becker two years his junior. 'I was walking past this small building that they used for entertainment of the student body, who were very idle and bored most of the term,' said Fagen, 'and I heard what I assumed was Howlin' Wolf playing ... I walked in and there was Walter with this red Epiphone guitar.' They clicked immediately, both being shy, snarky smartasses obsessed with bookish cool.

As a boy, Fagen had been deeply into sci-fi, and was even a member of the Science Fiction Book Club. 'That was the golden age of science fiction; all the great writers were active then. I loved C. M. Kornbluth, A. E. van Vogt. I liked the guys who were really social satirists. A lot of these guys came out of the socialist movement of the thirties, and they had a very funny way of criticising society. I really learned a lot from them. Certainly [from] Alfred Bester. He was a New Yorker. His first novel, *The Demolished Man*, got the rapid flow of life in the city, which I think is still present. There's something about the flow of Alfred Bester's prose that I think affected the way Walter and I write lyrics.'

When Fagen was a teenager, cool was rare, cool was underground. Nowadays the very idea of being hip is so commodified – and so readily available – that it is simply a part of a lifestyle experience. Back then, in the days when you had to seek out culturally subversive writers and like-minded souls, being cool meant being part of a very small club. 'When Walter and I met, we had a constellation of enthusiasms, really: science-fiction, jazz, black humour, novels by Thomas Berger, Terry Southern, Philip Roth, Vladimir Nabokov, Kurt Vonnegut especially. That certainly influenced the lyric writing. We also liked comic songwriting, like Tom Lehrer. He was a piano player and songwriter who wrote these grim, funny songs [Exhibit A: "Poisoning Pigeons in the Park"]. And then we were both fans of Frank Zappa and the Fugs.' This interest in Zappa manifested itself on *Katy Lied* in 'Everyone's Gone to the Movies', a deliberately tawdry tale of stag films and frat parties. Fagen was also a huge fan of W. C. Fields, a man who understood that 'most of life is just . . . you have to have the appearance that you know what you're doing'.

Becker and Fagen started writing together and eventually – after deciding to pursue songwriting as a career when they left college – spent months pestering the publishing teams in New York's Brill Building before being hired almost on a whim as staff songwriters by ABC Records producer Gary Katz and shipped out to California. Having initially tried to form various groups – with traditionally clever-clogs names such as Leather Canary and the Bad Rock Group (at one point employing fellow student Chevy Chase as drummer) – they realised their forte was writing, not performing. Then, finding their songs were unsuitable for any of ABC's artists – why would the likes of Dusty Springfield want to sing spiteful, gloomy songs about goofballs, druggy hipsters and lovesick aliens with macrocephalic heads? – they decided to form a band, building a musical edifice around themselves consisting of the finest studio musicians they could find.

When they first started looking for talent, they answered an ad seeking musicians: 'No assholes need apply'. And as Becker and Fagen didn't think they were assholes, they got in touch. They started recruiting like-minded virtuosos and eventually came up with a band hired for their musical ability rather than any notions of cool (which Fagen and Becker were convinced they both had in spades anyway). And so two droll, East Coast jazz buffs became responsible for creating one of the seminal West Coast rock bands of the early seventies, an ever-expanding group who would produce some of the decade's most important albums. From 1972's *Can't Buy A Thrill* and 1973's *Countdown To Ecstasy*, to 1980's *Gaucho*, Steely Dan perfectly fused West Coast cool with East Coast cynicism (as was once quipped, they were the Eagles as fashioned by Woody Allen). And they got away with it: their records sold in their millions, even though the pair of them seemed to despise the very people they were appealing to.

Musically, they favoured unnatural key changes, roller-coaster twists and jazz-driven hooks, while their songs were sardonic, sour and full of rather cruel wit. They were the smartass eggheads of rock, treating Chandleresque or sci-fi scenarios with sophomore black humour and, while it was easy to imagine yourself soaking up Steely Dan on drivetime FM radio in an open-topped Italian sports car making its way southwards at dusk on the Pacific Coast Highway, the band hated the very idea of such an image. They were sophisticated because the songs demanded it, not because of any ancillary lifestyle.

As the *New Yorker* put it: 'The lyrics were generally jaded assessments of young women, the older men who coveted them, and other humans caught at their least flattering moments.' Even their moniker was sarcastic, being the name of a dildo in William Burroughs's *The Naked Lunch*. Highly metropolitan, they excelled at manipulated isolation, while Fagen and Becker were labelled the most cynical and ferociously intelligent songwriters in the business.

'Our music is somehow a little too cheesy at times and turns off the rock intelligentsia for the most part,' said Fagen around the time of *Katy Lied*. 'At other times it's too bizarre to be appreciated by anybody.'

Experts said they welded jazz and rock into an alloy so smooth and shiny it was difficult to tell where the one ended and the other began, sneering at the world from a position of bohemian superiority so rarefied it was hard to tell exactly where it was situated. 'We were interested in a kind of hybrid music that included all the music we'd ever listened to,' said Fagen later. 'So there was always a lot of TV music and things in there. It was very eclectic, and it used to make us laugh: we knew something was good if we would really laugh at it when we played it back. We liked the sort of faux-luxe sound of the fifties, there was just something very funny about it. I grew up in a faux-luxe household and it was a very alienating world, so for me it has the opposite effect: muzak is supposed to relax you, but it makes me very anxious. So in a way, I think I get it out of me by putting some of it in my songs. Then I start to laugh at it when I hear it.'

Steely Dan were never very good at interviews, or at least couldn't be bothered to hide their disdain for music journalists. During one such encounter, Becker said to the unsuspecting hack: 'This is beginning to remind me of the joke where the guy from Oklahoma goes up to a New York cabbie and says, "Excuse me, could you tell me how I can get to Times Square, or should I just go fuck myself?"'

Fagen and Becker were far more radical than that and, although they expressed the same disdain for disco as they felt for the hegemony of mainstream rock, they enjoyed the fact that both were rebelling against the orthodoxy of FM radio. Not only that, but Fagen always seemed to be singing with one eyebrow raised.

Nevertheless, *Katy Lied* oozed a detached sophistication that was all its own, the highly polished surface disguising awkward time signatures and extra-credit guitar fills. 'We're actually accused of starting smooth jazz, which I don't think is exactly true,' said Fagen. Some people rejected the idea that a rock group could sound so slick.

For them, rock was 'guts and fire and feeling', in the words of Steely Dan fan Nick Hornby, 'not difficult chords and ironic detachment'.

Like a lot of those obsessed by recondite impulses, both Fagen and Becker were as intimidated as they were dismissive about the popular and the cool. At the time, Fagen said, 'We write the same way a writer of fiction would write. We're basically assuming the role of a character, and for that reason it may not sound personal.' Becker added, 'This is not the Lovin' Spoonful. It's not real good-time music.' White-hot chops and black humour, more like. Yet Steely Dan were actually cooler than anyone. Maybe not on a haberdashery level but cool all the same.

In a business that largely revolves around communication, both men took great delight in being as unengaged as possible. Fagen and Becker's forte was the intricate nature of their records, and they hated taking their band out on the road, which they saw as an endless litany of musical compromises. So eventually they did what the Beatles did: stopped touring and moved into the recording studio. At the time someone asked Fagen how they had managed it, and he said: 'Easy. We fired all the roadies so we couldn't go.'

The decision to become studio hermits was taken with *Katy Lied*, the first record they recorded with no intention of touring. As *Pitchfork*'s Mark Richardson wrote, in a retrospective piece on the album, 'Before *Katy Lied*, Steely Dan were a rock band, but this is the record where they became something else.'

It was at this stage in their career – midpoint, in fact – when the pair decided to pursue the kind of perfectionism that hitherto had been a by-product of the recording process rather than an end in itself. For instance, prog rock and the swatches of bands who practised it in the early seventies didn't demand virtuosity, but decreed complexity (often simply for the sake of it). *Katy Lied* was the album on which Steely Dan started to become fastidious about how their music was recorded. The notes on the back cover began: 'This is a high-fidelity recording. Steely Dan used a specially constructed twenty-four-channel tape recorder, a "State-of-the-Art" thirty-six-input computerised mixdown console, and some very expensive German microphones.'

Ironically, the recording process was actually quite traumatic. While Becker and Fagen were recording with producer Gary Katz and engineer Roger Nichols, they experimented with a new technology called dbx, which was meant to expand the dynamic range beyond the conventional limit of analogue tape. The system worked by compressing the signal on recording and then expanding it on playback, eradicating tape hiss in the process. But during the recording, something went very wrong.

'It was better sounding than anything you've ever heard to this date,' Katz told Cameron Crowe a couple of years later in *Rolling Stone*. 'Even *Aja* [1977 album]. Unbelievable. We went to mix it, and the tape sounded funny. We found out the dbx noise reduction system we were using was not functioning properly.' They obviously managed to salvage the album, but Becker and Fagen both said they could never listen to it again.

Lyrically, the album was peppered with a classic collection of Steely Dan characters, a motley crew of dishonourables who sounded as they had been invented by a cross between Damon Runyon, Cyra McFadden and Armistead Maupin. Cynical, sleazy, cryptic, satirical, these were words written by grown-ups for grown-ups, songs which were totally lacking in anything approaching empathy. This was journalism, plain and simple. 'The characters flailing clumsily throughout *Katy Lied* are paralysed by desires they aren't introspective enough to understand,' said Mark Richardson, 'so all they can do is keep stumbling forward.'

Of course, many critics – and, more importantly, many potential Steely Dan consumers – were put off by what they perceived as their relentless sarcasm. I fell in love with them when I was about fourteen, and most of my friends at the time were unconvinced at best. They thought Steely Dan were too clever for their own good, which, I suppose, has always been their USP, both good and bad. 'The words, while frequently not easy to get the definite drift of, are almost always intriguing and often witty,' John Mendelsohn wrote in a review of *Katy Lied* in *Rolling Stone*. 'Steely Dan's music continues to strike me essentially as exemplarily well-crafted and uncommonly intelligent schlock.'

Which is why we loved them so.

Katy Lied is the record that best epitomises the generic Steely Dan sound: jazz piano high up in the mix, deceptive cocktail bar arrangements, bird-like alto saxophone, faultlessly inventive drums (almost all played by the precocious twenty-year old Jeff Porcaro, who would soon become one of the world's most in-demand session players) and backing vocals courtesy of Michael McDonald, who would soon start to define not only the sound of the Doobie Brothers, but also the sound of yacht rock in general.

As the decade wore on, Becker and Fagen found themselves documenting the decadence and luxurious indolence of the West Coast experience, using the kind of artisanal skills that would have shamed a Renaissance painter.

'Where they once wrote about the delightfully sleazy underbelly of life in America from a remove, they started to write more about what they saw around them,' wrote Richardson. '*Katy Lied* is the fulcrum in this progression – it's messier, less sure of itself, besotted neither with youthful confidence nor veteran polish.'

'The Dan', as the music-paper fanboys would learn to call them, followed *Katy Lied* in 1976 with the equally kaleidoscopic *The Royal Scam* (which produced an unlikely novelty quasi-disco hit with 'The Fez'), with its own references to lifestyle, in the shape of 'Everything You Did', a jealous husband's rant having discovered his partner's infidelity. It includes the line, 'Turn up the Eagles, the neighbours are listening.' This was an obvious reference to the fact the Hollywood cowboys had become the most ubiquitous and most popular rock group of the decade, their gentrified Californian country rock exemplifying the way in which the sixties had been successfully repackaged for the seventies. Sure, the Eagles dressed as outlaws, but it was style in the same way Bruce Springsteen's authenticity was a style. There was a personal aspect to the reference, too. The band's Glenn Frey said, 'Apparently Walter Becker's girlfriend loved the Eagles, and played them all the time. I think it drove him nuts. So, the story goes that they were having a fight one day and that was the genesis of the line. The Eagles reciprocated, after a fashion, including a nod to Steely Dan in "Hotel California" by using the phrase "steely knives", a penile metaphor which alluded to the origins of their name.

In 1977 came Steely Dan's second masterpiece, *Aja*, the record that many musicians rate as the personification of musical excellence. Technically and sonically, *Aja* is beyond compare. (The late *New York Times* critic Robert Palmer said Steely Dan's music sounded like it had been 'recorded in a hospital ward'.) You rarely meet a musician who doesn't love some aspect of *Aja*, and whenever I've interviewed a rock star at their home, I've often seen a CD copy around the place somewhere. It used to be played constantly in those places where you went to buy expensive hi-fi equipment and can still be heard in the type of luxury retailers who understand the notion of immersive wealth. Having listened to the album's 'Deacon Blues', Ricky Ross named his band after it, while 'Peg' would become widely known because De La Soul sampled it on 'Eye Know'. Over thirty years after it was released it was deemed by the Library of Congress to be 'culturally, historically, or aesthetically important' and added to the United States National Recording Registry.

By 1977, Becker and Fagen had become such disciplinarians in the studio, they would hire dozens of session musicians to record the same guitar solo or drum fill until they felt they had something approaching what they had imagined. They became obsessive perfectionists who spent millions of dollars relentlessly torturing the dozens of grade-A guitarists who apparently weren't 'yacht-smooth' enough. Musicians would spend hours, sometimes days, in one of the many Los Angeles studios Steely Dan used to record *Aja*, only to find their work had been jettisoned in favour of someone else's.

'We just kept adjusting our standards higher and higher,' said Becker, 'so many days we'd make guys do thirty or forty takes and never listen to any of them again, because we knew none of them were any good; but we just kept hoping that somehow it was just going to miraculously get good.'

He later said: 'The studio is all about the idea of the setup, particularly for men. A room where you have all this technology to help you, and where you have some toys. It's about that space-age bachelor-pad vibe. The studio satisfies a lot of those urges. And you need air-conditioning, and a book with menus in it. It's kind of a minimum liveable standard, really.'

'Part of the Steely Dan fable is the exacting, often perplexing personal standards they demand of the performances of guest soloists,' said the journalist Chris Ingham. 'There is a scene in the *Classic Albums* TV programme on *Aja* where they revisit some of the accomplished-sounding but rejected guitar solos of "Peg", fading them up and down with terrifyingly dismissive comments. It's a fascinating glimpse of that select Club Of Two that has daunted so many musicians.'

At the time of *Aja*, Fagen and Becker were New Yorkers on location in LA and, although they revelled in the recording facilities and the abundance of great musicians, seemingly on tap – they spent their days getting studio tans as opposed to any other kind – they found the city increasingly ridiculous. 'LA was certainly a lot of laughs,' said Fagen. 'Neither of us really liked it, because we just weren't LA-type people. We called it Planet Stupid. Nobody seemed to understand us there.'

'Becker and Fagen are interesting characters, sort of isolationists by nature,' said one of their session musicians at the time. 'They live in these houses in Malibu, not near anybody, and I have a feeling LA helps them keep their music going on a certain level – they're almost laughing at the people in their songs.'

Almost?

Still, they weren't above sentimentality. There is always a kind of skeuomorphic feel about Steely Dan records, in that they are imbued with a certain nostalgia, even though the songs themselves are incredibly modern.

Aja was a case in point. Released at a time when both punk and disco were experiencing their own apotheoses, it seemed completely at odds with anything else. As a testament to that, the record was remixed thirteen times in the five months before its release.

When they returned to the studios in 1979 to record *Gaucho*, knowing *Aja* would be a hard act to follow, their obsessions got worse, exacerbated by Becker's substance abuse. Then, in January 1980, Becker's girlfriend died of a drugs overdose, causing him to withdraw from recording even further. Three months later, he was knocked down by a New York cab and hospitalised with fractures to his right leg.

None of this improved the duo's mood. On *Gaucho*, they were using up to six different rhythm sections for the same song. One of the small army of guitarists called in was Mark Knopfler of Dire Straits. He described the experience as 'like getting into a swimming pool with lead weights tied to your boots'. Toto drummer Jeff Porcaro, who played on the title track, said, 'From noon till six we'd play the tune over and over and over again, nailing each part. We'd go to dinner and come back and start recording. They made everybody play like their life depended on it. But they weren't gonna keep anything anyone else played that night, no matter how tight it was. All they were going for was the drum track.'

Fagen was such a neurotic perfectionist in the studio that people started to call him 'Mother'. (While recording his vocal for 'Home At Last' on *Aja*, he allegedly spent four whole days punching in the words 'Well, the' at the start of the chorus.) After the 250th mix of 'Babylon Sisters', the maintenance crew awarded him a 'platinum' floppy disk, hand-painted with silver nail polish. Ten days later, the production team were all back in the studio to attend to the second bass note in the second bar, which Mother had noticed was too tentative.

Steely Dan's lyrics have always been dispassionate, the songs labyrinthine and the guitar solos sarcastic. 'Years ago, I flew out to LA to visit a girlfriend who dumped me as soon as I arrived,' said producer and musician Mark Ronson, who is a massive Steely Dan fan. 'I couldn't change my ticket so I had to stay in LA, miserable, for five days. I bought the Steely Dan songbook and a cheap electric piano and stayed in my room for the duration of the time, teaching myself those songs. I don't often think of the girl but I use those amazing chord voicings nearly every day.'

Cyberpunk eminence William Gibson is a huge fan, and liberally sprinkles his novels with band references: 'A lot of people think of Steely Dan as the epitome of boring seventies stuff, never realising this is probably the most subversive material pop has ever thrown up.'

The band are a sampling smorgasbord, and have been grazed by Beyoncé ('Black Cow' on the J'Ty remix of 2004's 'Me, Myself and I'), Ice Cube ('Green Earrings' on 1992's 'Don't Trust 'Em'), Hit-Boy featuring John Legend ('The Boston Rag' on 2012's 'WyW'), Naughty by Nature ('Third World Man' on 1999's 'Live Or Die'), Kanye West ('Kid Charlemagne' on his 2007 single 'Champion') and dozens more. Somewhat perversely, Fagen and Becker were the winners of the 1999 award from the American Society of Composers, Authors and Publishers for the most-played rap song, 'Deja Vu (Uptown Baby)' by Lord Tariq and Peter Gunz, who had used the intro from 'Black Cow'.

'Like most bands from before my time, I discovered Steely Dan through rap music, specifically because "Peg" had been sampled by De La Soul on *3 Feet High And Rising*,' said Mark Ronson. 'That was years ago, and I discover new things every time I put a Steely Dan record on. I'm still discovering songs for the first time. No other band managed to let groove and intellect coexist as seamlessly. The most incredible rhythm sections with the most captivating narratives and these crazy chord changes.'

In 1975, those crazy chord changes were at odds with the rock orthodoxy of the time. Steely Dan weren't explicitly rock, they weren't prog and they weren't jazz. They were, almost by default, exception-ally particular. Although while we all thought there was some kind of grand design at play, they were adamant they were just doing what they wanted to do, regardless of who might like it, or indeed buy it.

'One thing people don't realise is we never, ever went for a specific kind of sound, even in the seventies,' said Fagen recently. 'It's really a function of what we like to hear. A kind of rhythm and blues foundation with jazz harmonies and my voice and a few other points of style give you that sound.'

A sound that confounded the ages then and continues to confound them now.

David Bailey Putting on the Ritz

5 December, Langan's Brasserie, Stratton Street, Mayfair

DAVID BAILEY LOVED THE IDEA of Ritz *but drew the line at having the magazine's logo stenciled on the side of his Rolls-Royce. 'Someone suggested he do that when we launched, but we were cool, not brash!'*

In 1975, the photographer was so sick of working for Vogue *that he wanted to launch his own title, something which wasn't redolent of the sixties, something which felt fresh. Essentially, he wanted a magazine he was proud of. '*Ritz *was the magazine I came up with, a stylish fashion and photography magazine that evoked the style of Fred Astaire,' he said.*

He initially wanted it to be like a newspaper, but the more he got into it the more he thought what he really wanted was a cross between Interview *and* Rolling Stone, *but very much for the British market. He'd been there at the birth of both magazines and thought a mix of the two would be perfect for London. He also wanted all the interviews to be strictly Q&A, without the journalist getting in the way. He wanted positivity, as he loved Andy Warhol, who was the most positive person he'd ever met. 'He loved everything. To extremes. Anyone who could turn a Campbell's soup can into a superstar had to be positive.'*

Bailey wanted that attitude and more art photography, more painters, more illustrators, more art school, more sex, more rock and roll. He certainly didn't envisage the magazine being full of paparazzi pictures of chinless wonders running around ripping each other's bras off. He'd been working for Vogue *for over fifteen years and felt he'd had enough, needing a raunchier magazine with more sex appeal. 'I wanted everything to be marvellous, as I was fed up with the English attitude of always knocking anybody and everybody. I liked the American can-do attitude, not the build-them-up, knock-them-down attitude of the English.' Bailey never understood why you would go to the bother of asking someone to be in the magazine, then tearing them to threads. Why put Jack or Mick or Elton on the cover if you were going to be bitchy about them inside?* Ritz *invented the British paparazzi, and Bailey wanted it to be full of photographs of people at parties, but not nasty gossip. Bailey enjoyed the silly, and the arch. One of his favourite interviews was the one he conducted with Orson Welles, who only said, 'No,' throughout the entire process. The other was with the Queen. 'Oh, hi!' said Bailey. 'Oh, hello,' said the Queen.*

Ritz *launched at the very end of 1975 as a monthly, featuring everyone from Bianca Jagger, Manolo Blahnik, Amanda Lear, Frank Zappa, Oliver Reed and Kraftwerk to Ossie Clark, Jasper Conran, Patti D'Arbanville, Sylvia Kristel, Antony Price and Jack Nicholson. One early issue had Bob Marley describing what hair products he used. The offices were deep in Covent Garden, right next to* The Lady, *around the corner from the restaurant Joe Allen, up a couple of flights of rickety stairs, and its contributors included Peter York, Craig Brown, Clive James and Gore Vidal ('He was a cunt, but a cunt with the biggest brain of anyone I ever met'). Bailey shot most of the pictures himself, but also used his friends Helmut Newton, Terence Donovan and Brian Duffy, and gave the paparazzo Richard Young his first break.*

Bailey was at the height of his belligerence and treated many of his assistants appallingly. 'The meanest thing I ever did to an assistant was in Paris when I told this little French girl that I'd run out of focus and asked her to go out and buy a bottle – a bottle of focus. She went all over Paris looking for it, the poor thing.'

When Bailey launched a trial issue, the newsagents WHSmith took some, he found out which shops they were in and went and bought them all – every issue he could lay his hands on. Consequently, they got a WHSmith contract.

The magazine either had their meetings in Langan's – lunches and dinners that Bailey always seemed to pay for – or else at his house in Gloucester Avenue, in Primrose Hill. He couldn't afford to hire studios, so shot a lot in a big room at his house that he'd turned into a studio. 'Langan's was our greasy spoon. All the Hollywood lot sat in the window seats – Jack Nicholson, Tony Curtis, John Travolta, Mick Jagger . . .' Langan's opened the same time Bailey launched Ritz, *so it was unsurprising it became the magazine's canteen. Peter Langan, who owned the restaurant with Michael Caine, was nearly as notorious as Bailey, and tales of Langan's escapades became the stuff of legend. Society columnist Nigel Dempster once reported in his column for the* Daily Mail: *'According to a well-known peer, who spent the night on the town with Langan, he received a phone call the next morning from him asking: "Were you still with me when I hit David Frost?"'*

Ritz *was influential for two reasons. Firstly, it invented celebrity publishing by bringing celebrity culture to the UK, although its imitators were soon to drag the concept down to the very gutter; and secondly it became a benchmark for independent publishing, encouraging the likes of* i-D, The Face *and* Blitz *to set up shop, those eighties inventions devoted to cataloguing their own social ecosystems. Bailey was as important to the seventies as he was to the sixties, as he understood how fame had become a commodity, not that he liked to analyse it. He would always say there was not any grand theory behind what he did, or indeed how he behaved. He'd say it was instinct, being in the right place at the right time, taking the right picture and not looking back. If the sixties had taught Bailey anything, it was the notion that you could bend culture by sheer force of nature. The seventies had taught him not to let it go.*

Ditch Black

To the untrained eye, the early seventies were good to Neil Young, as the likes of *Harvest* and *After the Goldrush* had made him a benign force in the post-Woodstock firmament. But then cocaine, Vietnam, Nixon and personal tragedy dragged him to the other side of the road. *Time Fades Away, On the Beach* and *Tonight's the Night* are considered by many to be the Rosetta Stone to understanding his entire body of work. Because of their dark, haunting brilliance, these albums are known as the 'Ditch Trilogy'.

Tonight's the Night by Neil Young

HER FIRST SCREEN APPEARANCE, IN 1969's *Easy Rider*, went uncredited and largely unnoticed. It was her second, a year later, in the comedy-drama *Diary of a Mad Housewife*, that made Carrie Snodgress a star. At the time, considered something of a feminist movie, all it really did was highlight spousal neglect, but it won her several awards, including a Golden Globe. It certainly had a huge effect on Neil Young, who became transfixed with its star. He eventually tracked her down to a local theatre where she was appearing and left a note on her dressing room table that simply said, 'Call Neil Young'. So, she did. By 1971, the couple were living together and in 1972 their son Zeke was born, and diagnosed with cerebral palsy. Three years later, the romance was largely over. 'He started hanging out with the guys, going to LA alone,' Snodgress said. 'Then one day he came back and said he thought it was time for me to leave.'

The relationship resulted in a stark, mostly acoustic album called *Homegrown.* His label at the time, Reprise Records, had accepted the

LP, finalised the cover art and set a release date. Then Young had a change of heart, 'as a lot of the songs had to do with me breaking up with my old lady,' he said. 'It was a little too personal . . . it scared me.' It shows how maladjusted Young was the time, thinking that *Tonight's The Night* could be interpreted as a more uplifting experience.

If *Blood on the Tracks* was therapy as narrative, Neil Young's brilliantly torpid *Tonight's The Night* was a grief dump, a record full of loss not just for his friends (Crazy Horse guitarist Danny Whitten and Young's roadie Bruce Berry had both died of drug overdoses in the months before the album's songs were written), but also for a decade. As *Rolling Stone* put it, 'the post-hippie, post-Vietnam demise of counterculture idealism, and a generation's long, slow trickle down the drain through drugs, violence and twisted sexuality'. Sounds like fun, huh? Well, it wasn't and wasn't meant to be. *Tonight's the Night* was an album not so much about death as about mourning. This was an extreme non-fiction album, a travelogue of seventies ennui that was darker than anything punk would ever dredge up. If the world had any interest in long-haired repudiation, here it was, in some painful glory. Included with the early original vinyl copies of the album was a less-than cryptic message written by Young: 'I'm sorry. You don't know these people. This means nothing to you.' As if to underscore the importance of Young's testimonial, he once claimed that a small package of glitter was included in initial pressings ('our Bowie statement'), spilling when the listener took the record out.

Young's three consecutive early seventies albums, *Time Fades Away* (October 1973), *On the Beach* (July 1974) and *Tonight's the Night* (June 1975) are considered by many fans the Rosetta Stone to understanding his entire body of work. Because of their dark, haunting brilliance, the albums are known as the 'Ditch Trilogy'. In the often quoted handwritten sleeve notes of his *Decade* compilation, which would come out in 1977, Young wrote, '"Heart of Gold" put me in the middle of the road. Travelling there soon became a bore so I headed for the ditch.' Didn't he just. The period is also referred to as the 'Doom' period or the 'Wilderness Years', and the impetus for the heart of darkness spiral can be traced back to *Harvest*'s 'The Needle and the Damage Done', Young's legendary anti-drug plea based on the heroin troubles of Danny Whitten.

These three albums found Young at an emotional low but an artistic high, grappling with inner turmoil, his extravagant and dangerous lifestyle, and his role as a burgeoning rock star. Ultimately, it would be a run of albums that would define Young's uncompromising and revolutionary spirit. *Tonight's the Night* wasn't a lot of fun, but then it wasn't meant to be. Young was always a Marmite artist, and there were as many who recoiled from his thin, whiney voice as those who found nearly everything that voice attempted (even the less successful stuff) fascinating. And it was always the Crazy Horse stuff that was most artistically satisfying, Crazy Horse who could make you feel as though you were sitting on the edge of the world, imported beer in hand, cigar stick in the other, watching all the madness below.

It was around this time that Young started the work that would lead him to be called the 'Godfather of Grunge' (sloppy drums, unreconstructed chords, plus lank hair, anti-designer jeans, trucker's T-shirts, sneakers), Young embracing his own guitar histrionics, subjecting his audiences to mammoth one-note solos and acres of feedback. 'It's not a macho display, like some bands have this strutting thing where they get up there and move around and they sweat and they pose,' he said. 'The sweating we do is because we're so far into it that we've forgotten how not to sweat. I start hyperventilating, my nose gets really cold, and I feel this cool breeze blowing in my face when it's about 110 on-stage. You just get to this point where nothing else is there, it's just all gone and you're taking off.'

When we describe the way music makes us feel, it's usually got something to do with abandon – feeling completely separate, cut-off, falling through the air, walking through the woods, flying way above everyone, standing on the cliffs looking at the midnight ocean, crouching in a cornfield and peering into the valley . . . Neil Young's searing guitar solos had the power to make us feel that way, especially when he regrouped Crazy Horse in the nineties; when an earthquake struck nearby San Francisco, on the last day of recording at Young's Woodside ranch, apparently no one noticed. The searing guitar on *Tonight's the Night* was a different kind of searing, although it was no less affecting. The record spoke to the dark shadow that Los Angeles had started to cast on his circle. Most of it was recorded on 26 August 1973, in Hollywood, while Watergate and the last gasps of Vietnam dominated television coverage; Charles Manson's death sentence had been commuted to life; Young had separated from Snodgress, and his world was full of drugs and paranoia.

Harvest had rubber-stamped Young in the popular imagination. It had been phenomenally successful, going platinum and becoming the best-selling album of 1972. In addition to changing Young's position in the marketplace, building on the commercial breakthrough of his work with Crosby, Stills & Nash, it had cast him as the male Carole King, a singer-songwriter no different from Cat Stevens or James Taylor. *Harvest* had turned the grumpy Canadian into something approaching a pop star. As summer turned to autumn in 1973, eighteen months after the release of *Harvest*, Young was twenty-seven years old. He was grappling with the inequities of fame while still spending an inordinate amount of time with a gang who were still acting like students, especially in their consumption of drink and drugs. And the excess was beginning to take its toll, on the singer as well as his friends. In August 1973, as Young started the sessions that would produce the bulk of *Tonight's the Night*, he found himself in the middle of a world he didn't especially like.

Tonight's the Night was recorded quickly – capturing the raw energy and spontaneity that Young would come to trust so much – mostly in a small rehearsal room behind a music equipment rental shop in Los Angeles in August and September 1973. Some of it was slapdash, but Young didn't really care; 'Borrowed Tune' was lifted from the Stones' 'Lady Jane', something he couldn't even be bothered to disguise in its lyrics. However, the album's raw sound and dark tone was initially rejected by Reprise, causing a two-year delay in its release. Its misery quotient was high.

'It was a very dark, intense but healing adventure,' said the guitarist Nils Lofgren, who played on the album. 'Neil wanted to make a live record, extremely rough, to not even have the musicians know the songs too well, the antithesis of production. We recorded in this pretty funky little room in Hollywood. We'd get together at dinner time, shoot pool, sip some tequila, pretty much till midnight, talking about Danny and Bruce, commiserating, and then get round a table and Neil would start showing us these songs.'

Tonight's the Night was a direct, no-filter, self-medicated expression of grief. Bruce Berry had been the brother of both Jan Berry of Jan & Dean fame and Ken Berry, owner of S.I.R. studios where the album was recorded. The tequila-sodden sessions were the first time the remaining members of Crazy Horse had played together since Danny Whitten's passing. The atmosphere Young created was that of a solitary figure riding through this album like a musical John Wesley Hardin: a rootless, renegade hippie (albeit one who wasn't as drug-addled as some of his colleagues). While the likes of Fleetwood Mac and the Eagles were spending millions on big-name producers, string sections, overdubs and choirs, Young recorded everything in one take, in a basement. They started playing at midnight and drove home just before dawn to their hotel every night for a month. Visitors came by late at night, they drank tequila and smoked weed. The photographer Joel Bernstein spent some time shooting the sessions, and said, 'It was like doing a documentary on nocturnal animals pulled out from under a rock; they looked like rodents when you shined a light in their eyes.' Young's own father, Scott, once described the album as 'a man on a binge at a wake', but even that doesn't do it justice. The record's inner label was black, unlike Reprise's trademark orange.

Towards the end of 1975 Young explained to Bud Scoppa of *Creem* magazine how the loss was both personal and professional: 'At S.I.R. we were playing, and these two cats who had been a close part of our unit – of our force and our energy – were both gone to junk. Both of them OD'd, and we're playing in a place where we're getting together to make up for what is gone and try to make ourselves stronger and continue.'

It has a rushed, haphazard vibe that you rarely get from a major record. The recording was spare and some have said the point when Young's voice cracks during 'Mellow My Mind' is perhaps the most poignant in his catalogue. 'Putting this album out is almost an experiment,' Young told *Rolling Stone*'s Cameron Crowe on release. 'I fully expect some of the most determinedly worst reviews I've ever had. I mean if anybody really wanted to let go, they could do it on this one. And undoubtedly a few people will.'

Having had the record rejected by his record company, Young then recorded a slightly more conventional work, *On the Beach*, but not before taking his rejected album on the road throughout Europe. He assumed the character of a sleazy Miami Beach MC, who excelled in Vegas-type one-liners, directed through a drink-soaked show. 'I slipped out of myself,' said Young. 'I might have been really slapping the audience in the face with some of the shows, but it was healthy for me. I was the same person for too long.' One thing that irritated Young was concertgoers continually calling out for him to play his more popular songs (in particular the ones from *Harvest* and *After the Goldrush*). On the UK tour that preceded the release of *Tonight's the Night*, he insisted on playing the album in its entirety. Every night. He liked to play both versions of the title track, so when the audience would get mad about not knowing songs and start shouting out, he would say, 'Here's one you've heard before!' They would start going nuts and then he would play 'Tonight's The Night' again, smiling behind his mirrored shades.

Young wasn't oblivious to its challenging nature, though, describing it as 'the first horror record'. Some critics hated it, as he had predicted. Writing in the *NME*, Steve Clarke said, 'This selection of songs does nothing whatsoever to enhance Young's credibility . . . We all know he can do better and even has better songs around. So why'd he release it? Is he really just laughing at us for taking him so seriously?' But, like many of Young's mid-seventies albums, *Tonight the Night* slowly grew in legend as the years went on, and is now considered one of his best.

Years later, Michael Bonner would interview Primal Scream's Bobby Gillespie about the record. 'It sounds absolutely fucking wiped out,' he said. 'The band sound drunk, but there's more to it than alcohol – they sound stoned and coked up. It's great the way it's almost slovenly. It's got the same kind of sleazy, rough, broken feel as *Exile on Main Street*. I like the subject matter of the songs, the way Neil Young was documenting the culture of the time, he was telling it like it was. It seems very truthful and naked, very emotional. It's almost as if you can feel the disgust in some of the songs. *Tonight's The Night* is him at his most descriptive, writing about the scene and the people of the time, in an incredibly truthful way. I love the weariness of the album, the darkness, the cynicism. That album was like escaping a lifestyle, certain people, a culture. It ends in death and darkness. Some people just never come out of that darkness, and some people do. Some people are just stuck there, and some people move on. It's a catharsis, I guess. It's not a celebration, he's taking stock, telling it like it is, saying, "This is what went down, I was there, but now I'm getting out of it." I think it's still as relevant today as it was in 1975.'

A few months after the album's release, King Crazy Horse finally gave his fanbase what they had been hankering after. Following the Ditch Trilogy came *Zuma*, an album that was immediately more pleasing and far more appealing. It contained hummable songs, shit-crazy guitar solos, terrific melodies and it wore a big smile. It even included one solid-gold classic, 'Cortez the Killer', one of the finest songs he'd ever recorded. Neil Young was finally back on the road.

The Naked Civil Servant

17 December, Thames Television, Teddington Studios

IT WAS PERHAPS NOT TOO much of a stretch to believe that Quentin Crisp had been a rent boy as a teenager. Nor was it difficult to believe he had already spent thirty years as a professional model for life classes in art colleges. After all, he was a flamboyant man with dyed hair who had walked along the edge of society. The interviews he gave about his unusual life attracted great curiosity, and he was sought after for his personal views on social manners and the cultivation of style. And what views they were. 'There is no need to do any housework at all. After the first four years the dirt doesn't get any worse,' was one rather famous one, as was, 'For flavour, instant sex will never supersede the stuff you have to peel and cook.'

Crisp was many things – an icon, pioneer, courageous, controversial – and his flamboyance ignited the third quarter of the twentieth century and paved the way for today's more tolerant, accepting and colourful world. His outlandish appearance – he sported bright make-up, dyed his long hair crimson, painted his fingernails and wore sandals to display his painted toenails – brought admiration and curiosity from some quarters, but generally attracted hostility and violence from strangers passing him in the streets. He attempted to join the army in 1939 when war broke out but was immediately rejected and declared exempt by the medical board on the grounds that he was 'suffering from sexual perversion'. He remained in London during the Blitz, stocked up on cosmetics, bought henna in bulk, and walked through the streets during the black-out, picking up GIs.

The Naked Civil Servant *was a 1975 television adaptation of Crisp's 1968 autobiography. It was published a year after the passing of the Sexual Offences Act, which partially decriminalised consensual sexual acts between men 'in private', aged twenty-one or over; it only sold 3,500 copies and largely went unnoticed. It was an unforgettable performance in one of the great TV dramas, said the* Guardian, *'all louche defiance, feline elegance, catty wit and understated loneliness. Crisp became a celebrity after the film, John Hurt became a star.' The audience in the UK were largely accepting, although the Americans were something else again. Controversy arose when the film first aired on US television, although it wasn't immediately apparent just how many protestors had actually watched the film they were complaining about.*

It didn't flinch from portraying the difficulties faced by gay men in the decades leading up to homosexuality's decriminalisation, but it presents them in keeping with the style of Crisp's own colourful storytelling. For many, watching the film for the first time was as shocking and as liberating as seeing David Bowie as Ziggy Stardust three years earlier. Crisp said Hurt played him perfectly, with pathos and sensitivity: 'He imitated my voice to perfection.' He identified as bohemian as much as queer and in his book traced a life in which London bedsits and rooming houses were home to those, like himself, whose deliberate social otherness meant a rejection of domestic mores. The difference with Crisp was how bold he was about his otherness; he didn't attempt to hide either his sexuality or his bohemian nature.

Crisp visited the set a few times, but basically left the filmmakers to get on with it. John Hurt said, 'Quentin was asked how camp we were allowed to be, and he said: "You can't be camp enough."' Crisp often referred to Hurt as 'my representative on Earth'. Until Crisp became notable himself, that is. 'He said that any film is at least better than real life'; this was one of the many ways he expressed his sincere admiration for the film that helped make him famous.

'I am a stereotype,' said Crisp (not his real name: he was born Denis Charles Pratt). 'I am an effeminate man. And when I was young, I and the whole world thought that all homosexuals were effeminate. And of course they're not. You can just see which people are effeminate; that's the only difference. So, I became a prototype of the effeminate man, because I was conspicuously effeminate. But camp is not something I do, it's something I am.'

He said he thought that him being gay was not remotely sexual. His problem was one of gender. He didn't want to be feminine in order to meet more men. Far from it, he wanted to be feminine to fulfil his idea of himself. People regarded him as having done the things he'd done in order to be different from other people. But he didn't want to be different; he wanted to be more like himself than nature had made him.

It was his uncompromising yet benign attitude that really clicked, while Hurt's impersonation made Crisp seem even more agreeable. The film didn't just make Hurt famous, it had a huge effect on Crisp, too, and when people realised he was as humble and as exact as Hurt had portrayed him, the public embraced him. At the time he was an anomaly; he wasn't strident or aggressive, and neither was he a Carry On caricature – not a Kenneth Williams or a Charles Hawtrey – instead he was a proud man who didn't accept compromise. He actually didn't think there was any compromise worth fighting for. Or giving up for. There were some who thought he acted and dressed the way he did simply to get noticed; in reality the opposite was true. Gay men on television at the time needed to be a cliché in order to survive; they needed to be family-friendly camp, a neutered stereotype that was easily digestible and therefore completely non-threatening. In fact the mediated view of gay men was so prescriptive it was almost offensive. With a little bit of help from John Hurt, Quentin Crisp changed all that. For ever.

The Marketing of Bob

Bob Marley's two concerts at the Lyceum in London on 17 and 18 July would become two of the most discussed, most fetishised gigs in modern music history. If you were there, you were *there*. Recorded using the Rolling Stones Mobile Studio, the resulting album – released in December to great fanfare and huge success – basically created the UK reggae market, making Marley a star in the process, as well as a folk hero, Rastafarian icon, freedom fighter and college-dorm poster boy.

Live! by Bob Marley and the Wailers

'ALL THE WAY FROM TRENCHTOWN, Jamaica – Bob Marley and the Wailers.' So began one of the uplifting albums of the decade. *Live!* was recorded on Friday 18 July 1975, at the second of two beyond-sold-out shows at the Lyceum Ballroom, London, and released in the UK on 5 December the same year. For the six months between these events, Londoners who were at the concert talked of little else, and reports were couched in terms of bemused wonderment.

Chris Salewicz, who was working for the *NME*, said it was nothing less than magical. 'The Lyceum was a revolutionary night,' he told me. 'It was significant.' Don Letts, who was running the clothing store Acme Attractions in the Kings Road (selling electric-blue zoot suits and jukeboxes and playing dub reggae) said it was the most exciting gig he's ever seen in his life, 'Like a religious experience'. The critic from *The Times* remarked on the 'curious odour' in the air, while another *NME* reviewer had his pocket picked.

The photographer Kate Simon, who shot Marley at both concerts, said they were shocking. 'The beauty of his voice, the brilliance of his band, the hypnotic power of the music. For me it was a calling to reggae. I wasn't prepared for it.'

Roger Steffens, the journalist, said, 'Outside was bedlam. There were people who were waiting in line almost all afternoon and the line was a couple of blocks long. And then there were these dreadlocked guys standing at the entrance saying, "Tickets! Tickets!" and people were just handing them their tickets. These guys would run back into the crowd and sell them. They were hustlers. The result was a near riot and they had to call out the fire department to spray people off the steps of the Lyceum.'

'Apart from the music, I can't remember much,' said another friend. 'I was unbelievably stoned. At least I think I was. The air was so thick with smoke that you didn't need to smoke anything yourself. All you had to do was breathe.'

What everyone agreed on was Marley's charismatic performance, the slick performance of the Wailers and the almost overwhelming sense of occasion. The reviewer from *Sounds* was so punch-drunk he compared it to Bob Dylan's infamous show at the Albert Hall in 1966. Marley's current tour would, he said, 'finally make reggae respectable'.

It was already moving in that direction.

'By the time Bob played the Lyceum, there was a definite buzz around reggae,' said Salewicz. 'It had probably started with the *Catch a Fire* album. I can't pretend I was that into it before then. I'd liked ska as a mod, but I wasn't listening to reggae in the early seventies until that album. There was an underground reggae vibe, but that was the catalyst. At the time, white rock fans really didn't like reggae, and they thought it was humorous, a bit of a joke. Then Virgin signed U-Roy, you had Toots and the Maytals, there was *Marcus Garvey* by Burning Spear and suddenly there was a buzz. Rock fans were very condescending about it, almost racist, actually. They called it music for thick people.' Beyond a novelty hit or two, cracking the international market remained a distant dream for reggae artists. Consequently, critics diminished it. According to them, reggae (or more accurately, ska) was only listened to by skinheads and children. Judge Dread, a white reggae singer, had UK hits with the lewd double-entendre singles 'Big Six', 'Big Seven' and 'Big Eight', which seemed to cancel out the finesse of classic ska singles such as Desmond Dekker & The Aces' 'Israelites' – which reached UK No. 1 in 1969 – and Dave & Ansel Collins' 'Double Barrel' two years later. In 1970, Chelsea football club began using Harry J Allstars' reggae instrumental 'Liquidator' as an official club song, although in the critics' eyes, this was further proof of the genre's unimportance. In one respect, the music's popularity was aided by rock's increasing self-seriousness. By the mid-seventies, reggae was still being treated as a curiosity by many, in much the same that 'world music' would be treated in the eighties.

The Lyceum concerts would change all that. For two nights, Marley would wear the mantle John Lennon had worn just over five years previously, when he had appeared wearing his famous white suit on the Lyceum stage with the Plastic Ono Band, as part of a benefit concert for UNICEF (the last time he would play in the UK). After these concerts, Marley would start to be deified, becoming one of those global rock stars who, like Lennon, seemed capable of administering some kind of universal cultural balm, his image on the walls of student halls of residence all over the world. He would be treated like a messenger.

Marley and the Wailers were touring to promote the *Natty Dread* album, which had been released the year before. Starting in Miami, Florida on 5 June, the band played much the same set list of songs across the US. 'By the time they got to London they were steaming,' says Salewicz. 'Olivia Harrison said she and George had been to see them play the Roxy in LA, and after the gig the first thing Bob said to George was, "Have you got any weed?" He said, "I was going to ask you that."'

The Marley shows had been feverishly awaited. Word of mouth had spread from his previous tours of the UK and while he had long been a hero among the Afro-Caribbean community, Marley was now on the hipster radar. On both nights, by early evening, long before the band was due to appear, the foyer was impassable and the queue stretched from under the portico into Covent Garden's hinterland. This was one of the first gigs that Marley's new promotor, Don Taylor, had booked for him. Taylor had a shrewd policy: he would always under-book his acts. When Marley could fill a place with five thousand people, he'd put him on in a nightclub and if he thought he could fill a twenty-five-thousand-capacity arena, he'd book him into a place that held ten thousand. This meant that there were invariably a lot of disappointed punters milling around outside the venues.

Inside the Lyceum, the excitement had started building long before 9.30 p.m., when the Radio London DJ Steve Barnard asked, rhetorically, 'Are you ready? Are you ready?' The crowd were glugging from cans and bottles, smoking and slowly dancing. Some were sitting down, waiting to be impressed. The roadies started to prowl about the stage. On one side, you could see a huge backdrop of Marcus Garvey, in ceremonial clothes. On the vast back wall there was a small picture, ringed with the red, yellow and green colours of the Ethiopian flag, of Haile Selassie, the embattled emperor of Ethiopia, Lion of Judah, considered by Rastafarians to be the godhead. And then Marley walked onstage, and the place went mad.

Having spoken to over two dozen people who went to the Lyceum concerts, there is still no consensus regarding the venue's retractable roof, which some remember, some don't. Neil Spencer, who was another journalist working for the *NME* at the time (and who would later go on to be editor), said the roof was one of the most defining things about the shows: 'The July weather was suitably tropical. Inside the Lyceum, the sweat ran in rivulets down the walls, despite the retractable roof being open. Up beyond, the night sky glimmered dimly beyond a haze of heat and smoke.'

'I saw Marley three times on the *Catch a Fire* tour,' says Salewicz. 'He did four nights at the Speakeasy, and I went to three of them. But the Lyceum gig was something else. He was sensational. And I mean, really sensational. My *NME* colleague Ian MacDonald had been the night before [it's difficult to find a staff member of the *NME* at the time who *didn't* go to these concerts] and he'd said how astonishing it was.

There were a lot of black people in the audience – it was about fifty-fifty, maybe sixty-forty white – but before the gig all the white people were sitting cross-legged on the floor. Some were even playing cards. They were still sitting on the floor when the set started. Everyone was very stoned, and it took a while for them to get on their feet. But all this stuff about the fire brigade being called to hose people down outside is bullshit. Although the Special Patrol Group [a serious crime division of the London Met] were there. They were keeping an eye on people, but I don't remember it being a particularly heavy vibe. All the white people there were hippies. The crowd were ready, expectant – it was like the coming of the messiah. You really had to hustle to get tickets. The *Natty Dread* album had been a Top 10 hit, and the people wanted to see what they were like.'

'The thing about the show is, it was largely attended by black people, and they had a hero,' said Kate Simon. 'And because the white music press was encouraged to go and check him out, they were in on it too, and he knocked everybody's socks off. Whether they were black or white. There was no pit like nowadays. Five years later he was playing Madison Square Garden, where nobody could get that close. At the Lyceum, people were on top of him and that made it great for a photographer because you'd see all the beautiful compositional opportunities of people with their fists raised, crowding round him.'

'This was an audience that had never seen itself before, an amalgam of white hipsters, recovering hippies, crisply dressed black youth and lofty Rastas,' wrote Spencer. 'It was a multi-culti crowd that would become familiar during the Rock Against Racism years, its ranks swollen by punk rock recruits, but here, on this steamy fecund night, was its inception.'

The myth of Marley had been very well managed by Chris Blackwell. Raised in Jamaica, Blackwell had started his label as a way of exporting the popular music he had grown up with. As soon as he met Marley, he knew he was special; he thought he had a certain something; he was small and slight but exceptionally good-looking and charismatic. When he signed Marley to Island Records, his idea was to try and break him like Jimi Hendrix or Sly and the Family Stone. Blackwell marketed Marley as a black rock act to white, college-educated rock fans and maturing hippies, who were drawn to reggae because they thought it was earthy and authentic. 'It didn't seem a weakening or a pasteurisation to market the Wailers as you would a rock act,' said Blackwell. 'There was no sellout. It seemed more compromising and condescending to continue to market reggae as niche music or some kind of exotica.' He sold him almost as a mythical figure, with a heavy emphasis on Rastafari and (obviously) ganja. He attempted to build intrigue and succeeded fantastically. The staff at Island thought he was crazy, as he'd poured at least a million into Marley, but he believed in him, and knew he could make him a star. 'No one knew much about Jamaica at the time,' says Salewicz. 'You had heard there was a bit of political turmoil but that was about it.'

Blackwell rearranged some of Marley's work on *Catch a Fire*, giving a studio sheen that he knew would appeal more to a rock audience. Marley travelled to London to supervise Blackwell's overdubbing of the album at Island Studios, which included tempering the mix from the bass-heavy sound, adding guitar solos and synthesiser to the album's final mix. On release, it was packaged like a rock record in a sleeve that looked like a giant Zippo lighter. Blackwell followed this with the album *Burnin'*, which included 'I Shot the Sheriff', famously covered by Eric Clapton.

Peter Tosh and Bunny Wailer then left Marley's band (Tosh saying he was fed up with what he thought of, unreasonably, as Blackwell's relentless 'fuckery') and in Jamaica the reaction was almost like the Beatles had split. They were the group who were putting the country on the world stage, but Marley managed to rise above it all, coming back with a new album, *Natty Dread*, credited to his newly reconfigured band, Bob Marley and the Wailers. The line-up of the Wailers that arrived in London in 1975 comprised the longstanding rhythm section of Aston Barrett (on bass) and Carlton Barrett (drums) together with Al Anderson (lead guitar), Tyrone Downie (keyboards), Alvin Patterson (percussion) and a depleted I-Threes vocal section of

Rita Marley and Marcia Griffiths (Judy Mowatt missed the London shows). And having just trekked all over the US, they were hot. They used the Rolling Stones Mobile Studio, a sixteen-track, state-of-the-art facility which had been used earlier year to record Led Zeppelin's *Physical Graffiti* and Bad Company's *Run with The Pack*. Parked in the road outside the venue, the Mobile faithfully captured the spirit of the occasion. The Lyceum had a terrible reputation for sound, but the Marley sound system was just perfect. Even the support act, Third World ('Uptown posh boys,' according to Salewicz), sounded good, a luxury rarely afforded to support acts.

One of the reasons the recording was so good was the man responsible for it – Blackwell's right-hand man, Danny Holloway – used microphones just to record the audience. 'My main contribution is that the crowd was recorded really well,' he told me. 'We had four microphones that were ten feet over the audience's head, and we caught magic. There were the Caribbean kids who were aware of Bob going back to the sixties and then there was the new white guys who were impressed by the media Bob was getting. So, it was interracial, and the chemistry was really strong between those two groups of people. There were also pickpockets at the show, which created another dynamic where you had to watch your bag and stuff like that. But it was generally a benign atmosphere. It was organic. There was a lot of anticipation, but you definitely felt like you were part of something. It was on another level. Bob and the band were really ready to deliver. The audience was so excited that they were almost like a participant in the show. Usually, you heard the audience through the vocal mics, so this was something different. I had to keep running back to the truck to make sure everything was working in the way it needed to be.'

Those who saw the Wailers' 1975 shows talk about them in awed terms, not least where 'No Woman No Cry' is concerned. Originally recorded for *Natty Dread*, by the time Marley performed it in London, it had become something almost talismanic, with the audience treating it like a Beatles song. From the moment the audience take up the chorus's refrain before the band do, it feels utterly magical, luminous even. A deeply spiritual song, it spoke volumes about life in the Trenchtown ghetto without saying anything explicit about injustice. Instead – warm-hearted and spiritual – it offered solace and hope and was a glorious celebration of life in the face of hardship. Infectious melody notwithstanding, it was still about tyranny – as rapper Chuck D would one day say, a battle cry for survival. Released as a single, it was a massive hit. The whole album was actually inspired by the song. Blackwell had seen Marley play the Roxy in LA and had been amazed at the crowd's reaction when he started playing it, singing along to it as though it were already an anthem. 'The best live albums are the ones where the audience sing along, so I thought, Boy, I'm going to make a live album! The Lyceum holds about two thousand people I think, and yet you'll find twenty thousand people who claim they saw one of them. But the whole idea really came from hearing people at the Roxy singing along with that song and hearing how the chord sequence sounded like a hymn. It just had magic.'

'It was like year zero, that concert,' said Don Letts, who counts the Lyceum concert he saw as one of the pivotal moments of his life. 'The next day we were empowered – we've seen Bob Marley!'

There was a press conference on the Friday afternoon following the Thursday night show and one journalist asked Marley if he had been scared by the crowd at the Lyceum. The slack security and the fact the venue didn't have any kind of pit meant there were times when the audience had almost been onstage. 'No,' said Marley emphatically, 'it no worry me so much. The only thing, I didn't want them pull me off the stage or hurt me. Them guy held me too hard. Them too strong, real big guys.'

It was one of the last times Marley would be so close to his audience. After the Lyceum concerts, and the release of *Live!*, he would never again be so available. By the end of the year, he was a star and the future of reggae had been cast.

Manhattan by the Ocean

31 December, Wilshire Boulevard, Beverly Hills

WHEN **NEW YORK** *MAGAZINE EDITOR Clay Felker was launching his Californian offshoot in 1975, he knew its premises needed to be impressive. So, he planned carefully:* New West's *offices would eventually be quartered in a modern, glass-and-steel building on Wilshire Boulevard in Beverly Hills, not too far from that truncated boulevard of retail heaven, Rodeo Drive. His original idea had been to open opposite the Beverly Hills Hotel ('where the New York literati stayed while in town'), but he soon realised this would involve opening an office on a highway. In the end, the Wilshire Boulevard offices were only a short walk from the Regent Beverly Wilshire, another chichi Beverly Hills hotel. The furnishings were bought off the set of* All the President's Men, *while the offices were designed to look like the editorial floor of the* Washington Post. *And when the* New York *editors he'd sent west for the launch needed cars, he rented a small fleet (thirty, in total), not Fords or Chevys but nifty little Alfa Romeos. If you were going to spend four million dollars on a magazine launch, you needed both your customers and your advertisers to see how ostentatiously you'd spent it.*

The fact that Felker – one of the most celebrated editors in New York – was launching a magazine on the West Coast showed how the balance of cultural power was changing. As much as anything it was a message to the East Coast media establishment that creative voices and cultural forces could be found outside the 212 area code. As a California clone of Felker's sassy, brassy New York *magazine,* New West, *it seemed, had*

327

only to come and be seen to conquer. The mix would be the same mix that worked on the East Coast: the sexy, service chic that worked so brilliantly in Manhattan, combined with snappy observations, tip-top fashion and political muckraking. Of course, Californians were a lot less intrigued by politics than New Yorkers, but they welcomed the option to change. New York *told readers where they could get the best bagels, olive oil, coffee beans and apartments. The city may have been falling apart, but* New York *was the style guide for the new conspicuous consumption. Felker thought he could do the same for the denizens of West Hollywood, Laurel Canyon and Santa Monica, those urban desperadoes who wanted their tennis whites whiter than their neighbours, who wanted the best arancini, the best avocados and the best lip surgery. Michael Wolff said Felker had the crass but revolutionary (revolutionary in the sense that it overthrew generations of class conceits) notion that we are what we buy. Especially in California.*

From the typeface and style of the cover to an article describing a search for purportedly the best fettuccine Alfredo in southern California, the new magazine followed the Felker formula to the letter. The first issue was fat, too, a whopping 172 pages, half of which were ads. There was a list of the seventy-six people shaping California's future (geddit?) and some do-it-yourself advice. 'Succulents are the sexiest plants: learn to love them,' said one headline. He thought of New York *as an extension of smart Upper East Side dinner parties, with peppery talk of politics, real estate, money, business and gossip; there was no reason this wasn't going to play in Bel Air or Pacific Palisades. His admirers liked to say he had observed something new happening in Los Angeles and had brought his own outsider's sense of romance and a fascination with power and status. They said his magazine had a new palette of interests, with no brow distinctions. Restaurants and tittle-tattle were as important as business, or politics.*

Felker had pioneered magazine journalism orientated to a single urban region, attempting to explore its peculiar problems and opportunities. With New West, *he had raised his sights to cover not just a city, but a complex region – at least nominally, the entire west.*

Predictably – and not without some accuracy – some said that his staff imported from New York were just 'carpetbaggers' who would write about Californians with the same kind of condescension that they employed back east, sophisticates patronising the people of Lotusland, with its relentless freeways, preoccupation with leisure pursuits and strange cults, and supposed lack of cultural bandwidth.

Felker was robust in his defence. 'We're not dumb enough to think you can put out a California magazine from New York,' he said. 'All of us are aware that attitudes are very different out here. What we're bringing is a curiosity, a competitive style; we can ask the questions, and local writers can provide the answers. In New York, the issue is the problems of the city and living in the city; in California, the issue is: How do you make the most out of life here?'

Trouble great and small crowded thick upon the magazine as soon as it launched, as Los Angelenos proved to be less insecure than the publishers expected. California was in the midst of a population boom and a pop-cultural explosion, and Felker was riding high from his success in Manhattan. Felker pledged to defy the popular notion of California as a land of 'fruit and nuts', saying, 'The endless chatter about smog, and cars and oddball cults and Western civilisation swallowed by Cowboy vulgarity' was 'thirty years old and pickled in aspic'. And for a while it worked, with writers as diverse as Tom Wolfe, Joan Didion and Joe Eszterhas appearing in its pages, kicking up dust and jabbing their fingers in people's chests, as well as their chakras. How could it fail? 'It's been the most successful launching of a new magazine there's ever been, both in circulation and advertising,' Daniel Mahan, the magazine's marketing director, crowed. Trouble brewed, though, and ownership soon passed to Rupert Murdoch. The magazine would soon be rebranded as California, *and the end of an era was nigh.*

Pre-Punk:
Goodbye California

ST MARTIN'S SCHOOL OF ART already had a storied history long before the dawn of punk, although the college's modern era really began on the evening of 6 November 1975, when the Sex Pistols played their first gig, supporting the pub rock and roll band Bazooka Joe on the fifth-floor common room of the art school in Charing Cross Road. Glen Matlock (the band's bassist) was studying there, and Sebastian Conran, the college's entertainment secretary, who would go on to design clothes for the Clash, helped the Pistols secure the gig (50p on the door!), which would turn out to be one of the most tumultuous debuts of all time. This was the gig where John Lydon famously wore baggy pinstripe trousers with braces and a ripped Pink Floyd T-shirt with 'I hate' scrawled over it. He looked like a young, fey Albert Steptoe masquerading as a soul boy. Bazooka Joe, on the other hand, looked not unlike many of the bands who would come in the Sex Pistols' wake. They had been formed by Danny Kleinman (who would go on to be a multi-award-winning advertising and movie director, responsible for the stunning title sequences for James Bond movies), Stuart Goddard (later to morph into Adam Ant) and John Ellis (who would find subsequent fame as a member of the Vibrators).

'We knew Glen Matlock, which I suppose is how we came to book them as our support band,' said Kleinman. 'We were doing a lot of gigs at that time, and we were playing all over the place – art colleges, universities, parties, and residences in pubs around London. We played to a lot of half-full rooms – that's how good we were. This one didn't seem to be different to any of our other gigs, but it did have quite a few memorable things about it. We were all art-school boys, thrashing away at our guitars, playing rock and roll, rockabilly, something edgier than the stadium rock of the time.

'We were slightly younger than the other bands on the pub rock circuit and slightly more aggressive. Maybe Eddie and the Hot Rods were quite in your face as well, but it was hard-hitting stuff and there used to be quite a lot of fights at the gigs. Particularly we played at the Stapleton Hall Tavern in Crouch Hill once a week, sometimes twice a week, and the locals didn't like it because it was quite noisy, we played loudly, they were always fights outside the pub and people were raucous when they went home. So, the pub manager asked me if I could do anything, so when we finished the gig I asked everyone to be respectful of the neighbours when you leave, don't make too much noise and go home quietly . . . and here's our encore, "Someone's Going to Get Their Head Kicked in Tonight". And the place would erupt with cans and bottles being thrown . . . There were a lot of ageing teddy boys who didn't like you playing songs unless they were exactly like the original, nor did they like you playing any new material, even if it was within the genre. We had a lot of younger teds who actually had enough hair to have quiffs, but if you saw a Ted without any hair you thought, Hang on, there's a bottle coming here.

'All the money we'd earned from playing in pubs, we'd put towards buying our own equipment. We'd bought all these 4 x 12 cabinets, 150-watt amps and a PA system and whatnot and we were still paying for it on the HP [hire purchase]. So we got to the gig, and the lifts were broken so we had to carry all this stuff up six flights of stairs, huge cabinets with concrete bottoms. It was a real pain in the arse. We set it up, got it all going, and then the support act turned up and said they hadn't got any equipment and could they borrow our PA, speakers and amplifiers and stuff? And we said, "Sure." I remember not being terribly impressed by the gig. I think they played a couple of cover versions [the

Who's "Substitute" and the Small Faces' "Whatcha Gonna Do About It"] and there was some not very competent playing. They played four or five songs and seemed OK until they started smashing up our equipment. Johnny Rotten started kicking the speaker cabinet, which we hadn't even finished paying for and had carried up all these flights of stairs and had just leant them out of goodwill, I thought, You're not Pete Townsend mate. So, I ran in and manhandled him a bit. I'm not a particularly violent person, but it wasn't my finest hour. I think Stuart was quite impressed by their performance, although I think I was too dull to realise this was a game-changing event. Fairly soon after this he left Bazooka Joe and reinvented himself as Adam Ant.'

'The Sex Pistols played in the student union, and it was a complete disaster,' said Stephen Jones, a St Martin's student who would go on to become a famous milliner. 'They turned up really late, played about two or three songs and I think either stormed off stage or the equipment broke down or something. And all the students wanted their money back and it was actually more of a scene about everybody wanting their money back than the fact that the Pistols played there.'

There were barely thirty people at the concert, and yet the demographic crossover had already started, with the St Martin's students offering the kind of sophisticated audience the Pistols would cultivate for their first year in business. At their early gigs there was always a smattering of soul boys (just as there was always an enthusiastic hippie contingent), as many bleached jeans and ski jumpers as there were torn T-shirts and plastic trousers. After that, the audiences started to become more institutionalised, not least because they started to ape generic rock and roll styles – one of the reasons that traditional rock fans found the original 1976 punks so disturbing is because they looked like demented soul boys rather than US-style rockers a la Lou Reed.

'There was a kid on my council estate called Stephen Marshall - who ended up being a roadie for Spandau Ballet – and Stephen was a big guy, a quite tough sort of guy,' said Robert Elms, the author and DJ. 'One day in November 1975 I saw him on my way to the tube

333

and previously he'd been this very normal straight guy. Suddenly I saw him walking alone and he's got a safety pin in his ear, and he'd shaved his eyebrows and he's really looking extreme, and I asked him what happened. And he said, "I went to see this group called the Sex Pistols." He said, "They're fucking terrible and fucking amazing." I remember thinking, Well, how does that work?'

Jah Wobble, who would go on to play bass in Public Image Ltd, and who was a good friend of Lydon, was there right at the beginning. 'It started in 1975, and it was great fun,' he said. 'It was weird, as you had all the gays and the mavericks and the girls and it was just great fun, before all the beer boys and the violence came in. Johnny Lydon was at Kingsway [college] with me, but then he disappeared with Sid. He turned up one day and I said, "Where have you been?" and he said, "I'm in a band," with a slight edge of aggression. Nobody was in groups then, and he may as well have said he was training to be a 747 pilot or a psychiatrist. He asked me if I wanted to come and see them rehearse, and it must have really meant something to him. I realised that this was a probably quite a big deal for him, because although he's got lots of front, he's actually quite a sensitive bloke. So, I knew I had to behave. I knew it was going to be shit. How could it not be terrible? But when I got there, I couldn't believe how good it was. And the musician who stood out the most was Glen Matlock. By a mile. He's quite a taciturn bloke, but what a player.'

Punk was a confluence of two opposing forces, namely pub rock and disco, to be crass, spurred on by those obsessives who worshipped at the bespoke platforms of Bryan Ferry and David Bowie. Already adored by the rock fraternity, Bowie and Ferry became revered by the dancers who frequented the standout southern clubs (such as the Goldmine in Canvey Island, the Royalty in Southgate and the Lacy Lady in Ilford), when they both adopted the rigours of the dancefloor – Ferry with 'Love Is the Drug' and 'Both Ends Burning' from *Siren*, and Bowie with *Young Americans*, two of the seminal records of 1975. Long before such things became pejoratives, Bowie and Ferry were seen as arbiters of cool, *indicators,* men who understood the benefits of looking sharp in both a retro and a modern sense, and who had purposefully rejected the ideological orthodoxies of the world of rock. Neither Bowie nor Ferry wore denim.

Kim Bowen, who would become one of the original Blitz Kids, was living at home in Farnborough in Hampshire when she heard about the Lacy Lady. 'My friend and I lied to our parents, saying we were staying at someone's granny's. We used to jump a train to Waterloo station, then dress up like prostitutes in the station because you couldn't leave your house dressed like that. My father – who was a plumber – would have killed me. Punk was just starting and we were wearing fishnets and leather. The Lacy Lady was an absolutely shitty, ugly-looking pub but it was miraculous because there would be all these cute guys who were really good dancers, dressed in clothes from SEX, like giant mohair sweaters and pegged pants and things.'

Chris Sullivan, another Blitz Kid, was a soul boy, who used to go to his local clubs in Merthyr Tydfil, as well as places like Wigan Casino, 'because of course there was no alcohol there, so you didn't have to be eighteen. I went to weekenders in Bournemouth, went to the Lacy Lady in Ilford. Then I came up to London in the September of 1975, when I discovered SEX and the King's Road. It was like, wow. This is amazing. The scene was really working-class. Joe Strummer's father might have been a diplomat, and Malcolm McLaren might have come from a rich family, but most people were working-class.'

At the time, Chris Salewicz was a journalist with the *NME*. 'You could feel something was in the air,' he said. 'Everything seemed unsatisfactory. The New York Dolls had already set the pace a few years earlier, in a way, because they pissed off people so much. I'd just started writing on *Let It Rock* magazine and everyone hated them. Critics and music fans were very reactionary, but you could feel something was coming. The answer wasn't Be-Bop Deluxe or Little Feat, and it was almost as though we were waiting for the seventies equivalent of Beatlemania. When it arrived, punk actually seemed quite logical. Ah, here it is.'

The changes were sociological as much as cultural. The *Time Out* street photographer Roger Perry spent the year photographing

London graffiti, most of which was political or surreal rather than tub-thumping or narcissistic, and which expressed a particular kind of social consciousness. Against a backdrop of urban decay, he captured the mood of a city in desperate need of change. The jazz singer and surrealist George Melly, who wrote the introduction to Perry's book, saw graffiti as collective liberation: 'In a world of supermarkets, office blocks, processed chickens, VAT forms, computers, ECT, time-and-motion studies, what graffiti proclaims is "Human beings rule OK!"' The messages are clear: 'Eat the rich', 'Fight back', 'I fought the law', 'Words do not mean *anything* today', 'Overcome apathy'. One stand-out is the situationist-inspired 'Same thing day after day – tube – work – dinner – work – tube – armchair – TV – sleep – tube – work – how much more?'

Five months before the St Martin's gig, on 5 June, Talking Heads had played their first gig, supporting the Ramones at New York's infamous downtown club, CBGBs. The Ramones had actually made their debut at the downtown club on 16 August 1974, but although they were already wearing matching black leather jackets and counting in all their songs by shouting '1-2-3-4', it wasn't until 1975 that a scene started to develop around what was starting to be called punk. May 1975 was also when Richard Hell (ex-Television) started the Heartbreakers, and recorded one of the first punk songs, 'Blank Generation', a classic rallying cry that encouraged Malcolm McLaren to ask the Sex Pistols to write their own version: 'Pretty Vacant'.

'In 1975, I'd just arrived in New York and was working as a cook at a film commune called Total Impact, a very strange place on 14th Street, where one day Legs McNeil came in for lunch,' said the Canadian-born journalist and filmmaker Mary Harron. 'I told him I wanted to be a writer and he said a friend of his, John Holmstrom, was starting a magazine called *Punk*. I thought it was a genius name, as before that people had just used the phrase, "You punk kid." One evening they came to get me from the office, where I was washing the floor, and they took me to CBGBs, to see the Ramones. It was the first time I'd been there.'

The Ramones were all from Forest Hills. Guitarist Johnny Cummings and drummer Tommy Erdelyi had been in a high school group called Tangerine Puppets. After they folded, they recruited Douglas Colvin and Jeffrey Hyman, and the Ramones were born. Colvin was the first to adopt the name 'Ramone', calling himself Dee Dee. He was inspired by Paul McCartney's use of the pseudonym 'Paul Ramon' when he stayed in hotels. Hyman then became Joey, and they were on their way.

Like a lot of early punks, they were really into nostalgia. 'Rock and roll was really washed up by the end of the sixties,' said Joey. 'People lost sight of what it was all about, and at that point it was all, you know, Emerson, Lake & Palmer, Pink Floyd and all these people. The most memorable songs were three minutes, whether it be the Beatles or the Stones or the Kinks or the Who or whomever.'

On 5 June, Talking Heads were still a three-piece (Jerry Harrison would join the following year), and it was the first time they were properly Talking Heads. David Byrne said it took them a while to settle on the name. Before that, he would change the band's name constantly, painting the different names on cardboard circles to put on Chris Frantz's bass drum. They settled on 'Talking Heads' after seeing the phrase in an issue of *TV Guide*. That night they played an early version of their signature tune, 'Psycho Killer', although Byrne was still playing guitar as though he were in the Searchers or the Swinging Blue Jeans. Afterwards, Johnny Ramone told them they sucked, although he already knew that; he had seen their audition and allowed them to support his band as he thought they'd make the Ramones look good.

In the next twenty-four months, the entire musical landscape changed completely. Expertise, virtuosity, innovation, experience and ambition were outlawed. Almost immediately. People became suspicious of creativity. The orthodoxy of purpose had changed completely: why build something when you could break it instead? Wouldn't that be more interesting? Wouldn't that be more fun? Wouldn't it look better? Music started to mirror the febrile atmosphere of the city, as well as the urgency of youth. Punk was plangent. Fast. Deliberately disposable. If, in 1975, the rock aristocracy had been on the verge of ascending to some kind of pantheon of talents – a kind of music industry Champions League, no doubt with a beautifully appointed front office in a Case Study-style bungalow somewhere on the fringes of Brentwood or Bel Air – by the spring of 1977, they were a laughing stock, for no other reason than it was simply their turn. Privilege, social mobility, money – all the inevitable by-products of a successful career – were suddenly nothing but an embarrassment. Artistic endeavour was no longer considered a laudable goal. The rock fraternity suddenly looked over-ripe, exclusionary, old. Who cared if you were any good? And what does good look like anyway? Have a look at this – I just broke it!

A telling conversation was the one between Jimmy Page and the legendary beat writer William Burroughs, in early 1975. Afterwards, Burroughs said that Page had told him there was some serious resentment building against the 'big rock stars in England now. You can feel it in the street, with people spitting on his Rolls in the Kings Road. And Jagger and his wife are refused taxis in London when the older drivers see who they are. Last year, Mick was punched in the face by a customs agent in Paris. The rock stars are facing a lot of hostility. Many of the ones I've met seem rather . . . lonely, actually.'

Because the ramparts were destroyed so quickly, 1975 (and in fact the very idea of '1975'), and anyone associated with it, was swept away, banished, diminished, consigned to a pre-punk box that remained sealed until punk's own legacy began to look a little tarnished. In narrative metaphor it was buried under a Malibu beach house. Nowadays, the music made in the mid-seventies is revered for its sophistication, its aspiration and determination, and not least for its execution. The passage of time only enhances the power of early seventies' 'heritage', 'legacy' rock.

Was there a better year than 1975? Was there a more evergreen year? Cases have been made for 1963, 1966, 1971, and many more before or since. But no year witnessed the kind of unified alchemy as 1975, a golden year that peaked right before our eyes.

The Playlist (75 from '75)

In the Lands of the Lotus Eaters

'Kometenmelodie 2', Kraftwerk
In 1975, music didn't really sound like this. Soon, not even Kraftwerk would sound like this.

'Isi', Neu!
Motorik dancing for a world that didn't yet understand it.

'Fire', Ohio Players
Never was a fire truck siren so sexy.

'Pick Up The Pieces', Average White Band
The producer Arif Mardin apparently argued against its release as a single because the song was a 'funk instrumental played by Scotsmen with no lyrics other than a shout'.

'For the Love of You', Isley Brothers
In its first month of release, this sold 500,000 copies, perhaps as the record sounded post-coital.

'Do It Any Way You Wanna', People's Choice
A classic that would be sampled by Dillinger, Serge Gainsbourg, Public Enemy and the Jungle Brothers.

'Thunder Road', Bruce Springsteen
Rock and roll finally goes widescreen.

'How Many Friends', The Who
Maturity in a boiler suit.

'Feel Like Makin' Love', Bad Company
This is what a power ballad sounded like in '75.

'Chez Maximes', Andrew Matheson & the Brats
Coulda, Shoulda, Woulda . . . Recorded in 1972-73 but not released until 1975, the Brats were like a UK version of the New York Dolls, and with a luckier role of the dice could have been as big as Mott the Hoople or even the Sex Pistols. The album this came from, Grown Up Wrong, *apparently only sold 563 copies on its initial release, although unlike everyone who heard the first Velvet Underground, almost none of the people who bought it went off to form a band. Matheson wrote a brilliant memoir,* Sick on You, *which should be required reading for anyone interested in the formative post-glam years of punk.*

'Tush', ZZ Top
As their guitarist Billy Gibbons said, 'It's that secret blues language – saying it without saying it.'

'Roxette', Dr. Feelgood
A pub rock classic that anticipated punk by at least eighteen months.

'They Just Can't Stop It (Games People Play)', The Detroit Spinners
Pervis Jackson was one of the singers on this, whose bass voice was responsible for '12.45', which for a while became his nickname.

'Rhiannon', Fleetwood Mac
The quintessential sound of '75. Stevie Nicks's performance of the song at the time was so impassioned that Mick Fleetwood said, 'Her "Rhiannon" in those days was like an exorcism.'

'Miracles', Jefferson Starship
So good it could have been by Fleetwood Mac.

'Any World (That I'm Welcome To)', Steely Dan
Steely Dan wrote difficult novels with great choruses.

'Jungle Waterfall', Return to Forever
Fusion from the right side of the tracks.

'Low Rider', War
Everything this band did was better than it needed to be.

'Mr Wilson', John Cale
The Velvet Underground meet the Beach Boys.

'My Little Town', Simon & Garfunkel
Simon & Garfunkel meet John Irving.

THE PLAYLIST (75 FROM '75)

'In France They Kiss on Main Street', Joni Mitchell
The apex of studio-produced poetry.

'Coney Island Baby', Lou Reed
Quite possibly his best song of the decade, and a world away from Metal Machine Music *(also released this year).*

'Both Ends Burning', Roxy Music
Bryan Ferry meets his people on the dancefloor.

'Goin' Back', Nils Lofgren
Making the Goffin King song his own.

'Young Americans', David Bowie
Off to the disco for a volte-face.

'Quiet Storm', Smokey Robinson
And so, this is foreplay.

'I Love Music', The O'Jays
The feel-good party anthem of the season.

'Shining Star', Earth, Wind & Fire
Producer Sandy Pearlman erroneously called Earth Wind & Fire 'the closest thing to a black heavy metal band', while this was a harbinger of all the success that was to come.

'Billy Jack', Curtis Mayfield
Gun crime prescience.

'Shame, Shame, Shame', Shirley & Company
Produced by Sylvia Robinson, who would go on to create Sugar Hill Records.

'Flying Junk', 10cc
A Swinging Sixties evocation.

'Jackie Blue', Ozark Mountain Daredevils
Bill Mann of the Montreal Gazette *assumed the track was sung by a woman, dismissing it was 'an outrageous [knockoff] of Fleetwood [Mac]'s sound, down to the female lead.'*

'How Long', Ace
It came out of the pubs and ended up – after a polish and a brush-up – on FM radios all over the US.

'You're My Best Friend', Queen
Producer Roy Thomas Baker making British royalty sound American.

'The Trip', Bobbi Humphrey
It's not 'Harlem River Drive', but it's still the best Bobbi Humphrey record of the year.

'You're No Good', Linda Ronstadt
In the previous twelve years it had been a hit for Dee Dee Warwick, Betty Everett and the Swinging Blue Jeans, but Ronstadt's super-swampy version went all the way to No.1.

'Love to Love You Baby', Donna Summer
It's impossible to exaggerate just how influential this record was, at whatever length.

'The Köln Concert Part 1', Keith Jarrett
Replacing Miles Davis's A Kind of Blue *as a dinner-party staple.*

'Mister Magic', Grover Washington
Who ate a bowl of funk.

'When an Old Cricketer Leaves the Crease', Roy Harper
Playing, and beating Ray Davies at his own game.

'Tangled up in Blue', Bob Dylan
A mobile marriage advice bureau.

'At Seventeen', Janis Ian
Heartbreaking to hear, and probably more so to write.

'Enjoy It', Brian Protheroe
Essentially English, and therefore not necessarily successful.

'The Hustle', Van McCoy
Watch out for the piccolo tootle.

'Winter in America', Gil Scott-Heron and Brian Jackson
Telling it like it is.

'Make Me Smile (Come Up and See Me)', Steve Harley & Cockney Rebel
A glam record built for American radio.

'We All Fall in Love Sometimes', Elton John
A song about the relationship between Elton and Bernie Taupin. 'It made me well up because it was true,' he said. 'I wasn't in love with Bernie physically, but I loved him like a brother. He'd was the best friend I'd ever had.'

'Lovin' You', Minnie Riperton
Produced by Stevie Wonder who, to avoid contract conflicts, was credited as 'El Toro Negro', Spanish for 'Black Bull', as Wonder's star sign is Taurus.

THE PLAYLIST (75 FROM '75)

'Loving Arms', Millie Jackson
Unusual in that it was recorded in a studio to sound like a live recording, with Jackson giving a spoken intro over a tinkling piano and audience chatter, with rousing applause at the end.

'Trenchtown Rock', Bob Marley
As live as you were ever going to need it, from the Lyceum recording.

'Marcus Garvey', Burning Spear
Remixed with great success by Chris Blackwell, who believed the original Jamaican mix to be too threatening, or at least uncommercial, for a white audience.

'Redondo Beach', Patti Smith
Fountainhead proto-punk poet makes uncharacteristically upbeat reggae classic.

'Spare Parts 1 (A Nocturnal Emission)', Tom Waits
Recorded over four sessions in July in the Los Angeles Record Plant studio in front of a small, invited audience set up to recreate the atmosphere of a smoky jazz club.

'Kashmir', Led Zeppelin
A spiritual travelogue, with massive drums.

'Idiot Wind', Bob Dylan
The best Dylan song of the decade.

'Wake Up Everybody', Harold Melvin & The Blue Notes
When the Sound of Philadelphia was briefly the sound of America.

'White Punks On Dope', The Tubes
The Rocky Horror Show *reborn in San Francisco.*

'Movin'', Brass Construction
Massive.

'Why Did You Do It', Stretch
Basically, a song by a band complaining about being a Fleetwood Mac tribute act.

'Lady Marmalade', Labelle
Well, would you?

'Just Another High', Roxy Music
If Bob Dylan went to nightclubs.

'Borrowed Tune', Neil Young
On this, the Godfather of Grunge sounds disorientated, blind drunk, and boasts about how he's stolen the melody from the Rolling Stones.

'Long Distance Love', Little Feat
A song so moving it made an NME *writer (Peter Erskine) cry, alarmingly enough.*

'99 Miles from LA', Albert Hammond
An homage to Jimmy Webb in all the best ways.

'Boulder to Birmingham', Emmylou Harris
Recounts her feelings of grief in the years following the death of her mentor, Gram Parsons.

'Melody', Rolling Stones
A great song on an almost song-free album.

'Dimming of the Day', Richard & Linda Thompson
Having adopted the Sufi faith the previous year, and subsequently moving into a commune in London, the Thompsons flexed their creative muscle through songs which reflected their newfound faith.

'Song for Ché', Robert Wyatt
Written by Charlie Haden, an original member of the ground-breaking Ornette Coleman Quartet.

'Blue, Red and Grey', The Who
Plaintive, tender, adult.

'You Go to My Head', Bryan Ferry
'These Foolish Things' Part Two.

'Mummy Doesn't Go to Parties Since Daddy Died', Sadistic Mika Band
Their name was a parody of the Plastic Ono Band, so what was not to like?

'Jesus' Blood Never Failed Me Yet', Gavin Bryars
Based on a loop of an unknown homeless man singing a brief improvised gospel stanza, released on Brian Eno's Obscure Records.

'Shine On You Crazy Diamond', Pink Floyd
Wish You Were Here *was the last great Floyd album, and this was its greatest song.*

'Another Green World', Brian Eno
Used as the theme music for the BBC Two arts series Arena, *this was the inevitable intersection of pop and culture.*

'Tomorrow Belongs to Me', The Sensational Alex Harvey Band
Musical theatre on the terraces.

Acknowledgements and Sources

THE PUNK EXPLOSION OF 1976 was so tumultuous, and in many respects so unforgiving, that the Year Zero ethos it espoused gave short shrift to much that came before it. Overnight it became fantastically uncool to have once liked Pink Floyd, or to have seen Thin Lizzy in concert (I was guilty on both fronts*), and forced many people to quickly cover the tracks of their musical taste. We all have bought records we've no doubt become embarrassed about, but in 1976 being caught with a Poco album or Gong bootleg (not guilty, I hasten to add) would have been enough to demonise you for ever. When I arrived at the Ralph West Hall of Residence in the late summer of 1977, I distinctly remember people hiding particular albums depending on who had just come into their room. Others were not so lucky. A friend of mine, Fiona Dealey, had an overly enthusiastic Genesis fan as her neighbour, who played *Trick of the Tail* so often ('They're my favourite group,' she'd tell anyone she met, whether they were listening or not), she asked if she could change her room (a request that was unfairly denied). I don't blame her. Fiona, an old soul girl whose idea of an instrumental break was an orchestral bridge on a Philadelphia International single, quite rightly described the noise she heard blaring through her walls each

---

* (As it was, in 1975 I spent most weeks going to the Nag's Head on London Road in High Wycombe, a bumfluffed fifteen-year-old watching everyone from Dr. Feelgood, Ducks Deluxe and the Kursaal Flyers to Sassafras, Kilburn and the High Roads and Eddie and the Hot Rods.)

night as twaddle. Elsewhere at the Ralph West, people disguised their record collections in the same way they disguised the way they spoke, dropping their aitches or watching their glottal stops, again depending on who they were with, or who had just walked into the room.

However, there were many records from the mid-seventies that I refused to bury, be they old soul singles (catch me in a particular light and I'll still say Esther Phillips's version of 'What a Difference a Day Makes' is my favourite record of 1975) or so-called 'difficult' albums I had venerated relentlessly, probably because I'd played them until I liked them (*An Electric Storm* by White Noise, for instance, a record I still listen to; released in 1969 but bought on holiday in a record shop in Torquay in the summer of 1975). Punk was a correction as much as anything, a much-needed swipe against pomposity, arrogance and age (oh, and flares), and its scattergun ideology didn't care who got shot. Its 'You're either on the bus or off the bus' approach naturally meant this new orthodoxy immediately banished the past, demonising a whole swathe of stuff that was now deemed Old Wave.

Fifty years on, it's plain to see that 1975 was a pretty good year for records, and for every bloated prog-rock 'Meisterwerk' and novelty pop single there was a *Horses* or a 'Games People Play' (the 'Detroit' Spinners were so called in the UK lest we confuse them with the undemanding Liverpudlian folk group of the same name). It was a year as rich and as varied as any before it, or after it, and for a while I have wanted to write a book to try and redeem it. Inspired by the likes of Jon Savage's *1966: the Year the Decade Exploded*, David Hepworth's *Never a Dull Moment: 1971, the Year that Rock Exploded*, and *Bill Bryson's One Summer: America, 1927*, as well as Ariel Leve and Robin Morgan's *1963: The Year of Revolution*, and *1956: The World in Revolt* by Simon Hall, I wanted to look at a non-pivotal year with new eyes, and actually old eyes.

In my quest I have been ably assisted by many people. I'd especially like to thank Jimmy Page, who spent several hours in his west London house last summer taking me through the meticulous assembly of *Physical Graffiti*; Chris Salewicz and Island Records' Danny Holloway, who relived their evenings watching Bob Marley at the Lyceum; Jah Wobble, who recalled his experiences of the early Sex Pistols concerts while sipping tea in the Chelsea Arts Club; Courtney Love, who told me about her love for Bob Dylan and *Blood on the Tracks* at her home in Notting Hill; Mark Ronson, who rhapsodised about Steely Dan; and Peter York, who carefully described his navigation of the social tributaries and West Wonderland hot spots of that particular summer (appropriately enough in the Groucho Club, which, when it opened in 1985, killed off the seventies for good). I spoke to many others about their recollections of 1975, and the mood music of the mid-seventies, and I'd like to thank David Bailey, Bob Colacello, Alan Edwards, Robert Elms (whose *Live! Why We Go Out* captures something few music books ever have), the exceptional Duggie Fields, Mary Harron, Nick Kent, Paul Smith, Paul Stolper, Chris Sullivan and Pete Townshend. I'd also like to thank David Bennington, Chris Blackwell, Michael Bonner, Michael Bracewell, Mark Cecil, Dan Chiasson, Steve Clarke, Elvis Costello, Winston Cook-Wilson, Fiona Dealey, Robin Derrick, Brian Eno, Bryan Ferry, Guy Fletcher, Corinna da Fonseca-Wolheim, Martin Fry, William Gibson, Paul Gorman, David Hepworth, Stephen Holden, Barbara Hulanicki, Chris Ingham, Keith Jarrett, Elton John, Stephen Jones, Terry Jones, Danny Kleinman, Don Letts, Nick Logan, Gerald Lyn Early, Andrew Male, Greil Marcus, Dave Marsh, Todd Mayfield, John Mendelsohn, Joni Mitchell, Suzanne Moore, Anthony O'Grady, Terry O'Neill, Tony Parsons, Antony Price, Marco Pirroni, David Ritz, Bud Scoppa, John Seabrook, Charles Shaar Murray, Kate Simon, Neil Spencer, Bruce Springsteen, Donna Summer, Quentin Tarantino, Graeme Thomson, the wonderful Keith Wainwright, Sarah Walter, Paul Weller, Richard Williams and David Yaffe. Particular thanks goes to Eric Harvey, for his fascinating *Pitchfork* piece on Quiet Storm, which helped me enormously. 'The inside story of Keith Jarrett's iconic Köln Concert' by Stuart Nicholson was also fascinating, as was John Lewis's piece in the *Guardian*, and Charles Waring's piece on *udiscovermusic.* And Mark Blake's brilliant *Classic Rock* romp, 'Cocaine, Quaaludes and chaos: the Casablanca Records story' opened many avenues for me. Other helpful articles are mentioned in the text itself. In the David Bowie chapter, I've used some material from interviews conducted for previous projects with Bowie, including Tony Visconti, Luther Vandross, Bob Harris, Ken Scott, Mike Garson, Ava Cherry, Carlos Alomar, Earl Slick, Gus Dudgeon and Bowie himself. Thank you also to *Atlanta Magazine*, BBC, *Blues & Soul*, *Creem*, the *Detroit News*, *Ebony*, *Far Out*, the *Guardian*, *Impetus*, the *LA Times*, *MOJO*, the Neil Young Archives, *NME*, the *New Yorker*, the *New York Times*, *The Paris Review*, *Pitchfork*, *The Quietus*, *Rolling Stone*, *Uncut*, the *Wall Street Journal* and the *Washington Post*. Every book listed in the bibliography was helpful to some degree, if only for suggesting why something might have happened in the first place.

Bibliography

Requiem for a Nun by William Faulkner, Random House, 1950

Couples by John Updike, Knopf, 1968

Rock Dreams by Nik Cohn and Guy Peellaert, Pan Books, 1973

Zen and the Art of Motorcycle Maintenance by Robert M. Pirsig, William Morrow, 1974

The Philosophy of Andy Warhol (From A to B & Back Again) by Andy Warhol, Harcourt Brace Jovanovich, 1975

The Painted Word by Tom Wolfe, Farrar, Straus & Giroux, 1975.

The Serial by Cyra McFadden, Apostrophe, 1977

Rolling Thunder Log Book by Sam Shepard, Viking Press, 1977

'I Gave Them a Sword': Behind the Scenes of the Nixon Interviews by David Frost, Morrow, 1978

Style Wars by Peter York, Sidgwick & Jackson, 1980

From A. to Biba by Barbara Hulanicki, Tyne & Wear Museums, 1984

Long Time Gone: The Autobiography of David Crosby by Carl Gottlieb and David Crosby, Doubleday, 1988

Holy Terror: Andy Warhol Close Up by Bob Colacello, HarperCollins, 1990

England's Dreaming by Jon Savage, Faber & Faber, 1991

Nova: 1965-1975 by David Gibbs, Pavilion, 1994

1975

You'll Never Make Love in This Town Again by Joanne Parrent,
    Robin Greer et al., Dove Books, 1995

Last Night a DJ Saved My Life: The History of the Disc Jockey
    by Bill Brewster and Frank Broughton, Headline, 1999

No Sleep Till Canvey Island: The Great Pub Rock Revolution
    by Will Birch, Virgin, 2000

Kraftwerk: I Was a Robot by Wolfgang Flür, Sanctuary, 2003

Chronicles: Volume 1 by Bob Dylan, Simon & Schuster, 2004

*Hotel California: Singer-Songwriters & Cocaine Cowboys in the L.A.
    Canyons* by Barney Hoskyns, Harper Collins, 2005

Re-make/Re-model: Becoming Roxy Music by Michael Bracewell,
    Faber & Faber, 2007

On Some Faraway Beach: The Life and Times of Brian Eno
    by David Sheppard, Orion, 2008

Crisis What Crisis? Britain in the 1970s by Alwyn W. Turner,
    Aurum Press, 2008

And Party Every Day: The Inside Story of Casablanca Records
    by Larry Harris, Backbeat Books, 2009

Canyon of Dreams: The Magic and the Music of Laurel Canyon
    by Harvey Kubernik, Sterling, 2009

When the Lights Went Out: Britain in the Seventies by Andy Beckett,
    Faber & Faber, 2009

Robert Altman: The Oral Biography by Mitchell Zuckoff,
    Alfred A. Knopf, 2009

Hot Stuff: Disco and the Remaking of American Culture by Alice Echols,
    Norton, 2010

Just Kids by Patti Smith, Ecco, 2010

State of Emergency: The Way We Were: Britain, 1970–1974
    by Dominic Sandbrook, Allen Lane, 2010

BIBLIOGRAPHY

In the Seventies: Adventures in the Counterculture by Barry Miles,
  Serpent's Tail, 2011

Disco: The Music, The Times, The Era by Johnny Morgan, Sterling, 2011

Going to Sea in a Sieve by Danny Baker, Weidenfeld & Nicolson, 2012

Who I Am by Pete Townshend, Harper Collins, 2012

Eminent Hipsters by Donald Fagen, Viking, 2013

Play On: Now, Then and Fleetwood Mac by Mick Fleetwood,
  Hodder & Stoughton, 2014

*Live from New York: the complete, uncensored history of Saturday Night
  Live as told by its stars, writers, and guests* by James Andrew Miller,
  Little, Brown, 2015

Going Into the City by Robert Christgau, Dey Street, 2015

David Bowie: A Life by Dylan Jones, Preface, 2017

Kraftwerk PUBLIKATION: A Biography by David Buckley,
  Omnibus Press, 2017

Face It by Debbie Harry, HarperCollins, 2019

Remain in Love: Talking Heads, Tom Tom Club and Tina by Chris Frantz,
  St. Martin's Press, 2020

Kraftwerk: Future Music from Germany by Uwe Schütte, Penguin, 2020

The Sound of the Machine: My Life in Kraftwerk and Beyond
  by Karl Bartos, Omnibus Press, 2022

*Once Upon a Time World: The Dark and Sparkling Story of the French
  Riviera* by Jonathan Miles, Simon and Schuster, 2023

Live! Why We Go Out by Robert Elms, Unbound, 2023

I Was There by Alan Edwards, Simon & Schuster, 2024

Neu Klang: The Definitive History of Krautrock by Christoph Dallach,
  Faber & Faber, 2024

Under a Rock: A Memoir by Chris Stein, Little, Brown, 2024

Index

INDEX